THE NATURE–NURTURE DEBATES

How is it possible that in more than 100 years, the nature–nurture debate has not come to a satisfactory resolution? The problem, Dale Goldhaber argues, lies not with the proposed answers, but with the question itself. In *The Nature–Nurture Debates,* Goldhaber reviews the four major perspectives on the issue – behavior genetics, environment, evolutionary psychology, and developmental systems theory – and shows that the classic reductionist strategies (behavior genetics and environmental approaches) are incapable of resolving the issue because each offers a false perspective on the process of human development. It is only through a synthesis of the two holistic perspectives of evolutionary psychology and developmental systems theory that we will be able to understand the nature of human behavior.

Dale Goldhaber is Associate Professor Emeritus at the University of Vermont. In 2010, he received the John Dewey Award for Excellence in Teaching from the College of Education and Social Services. Professor Goldhaber is the author of *Theories of Human Development: Integrative Perspectives* and *Life-Span Human Development* in addition to numerous articles related to human development and early childhood education.

9/13

The Nature–Nurture Debates

Bridging the Gap

DALE GOLDHABER

University of Vermont

CAMBRIDGE UNIVERSITY PRESS
Cambridge, New York, Melbourne, Madrid, Cape Town,
Singapore, São Paulo, Delhi, Mexico City

Cambridge University Press
32 Avenue of the Americas, New York, NY 10013-2473, USA

www.cambridge.org
Information on this title: www.cambridge.org/9780521148795

First published 2012
Reprinted 2013

Printed in the United States of America

A catalog record for this publication is available from the British Library.

Library of Congress Cataloging in Publication Data

Goldhaber, Dale.
The nature–nurture debates : bridging the gap / Dale Goldhaber, University of Vermont.
pages cm
Includes bibliographical references and index.
ISBN 978-0-521-19536-2 (hardback) – ISBN 978-0-521-14879-5 (paperback)
1. Nature and nurture. I. Title.
BF341.G56 2012
155.7–dc23 2012007501

ISBN 978-0-521-19536-2 Hardback
ISBN 978-0-521-14879-5 Paperback

To Mazy and Oli
Who remind me every day of the wonderful tapestry
that is the process of human development

Contents

Preface

Here is a riddle for you. What has been solved over and over again and yet still remains unsolved? The answer – the nature–nurture debate. Admittedly not a very good riddle, but nevertheless a true statement. The question is like one of those candles we used to have on our birthday cakes when we were children, the ones we would blow out and they would just light again and we would blow them out again and once more they would relight. Eventually we caught on that the problem wasn't with our blowing but with the candle. The focus of this book reflects the same sentiment. The problem isn't in the answer; it is in the question. We need some new candles to light.

The Nature–Nurture Debates: Bridging the Gap reviews both contemporary and historical approaches to the problem, from the perspective of both theory and method and the implications of each of these approaches. Without giving too much of the plot away, suffice it to say that the candle that needs replacing is the reductionist model. An approach that partitions variance into independent main effects will never resolve the debate because, by definition, it has no choice but to perpetuate it. The "new candle" is one consistent with the emerging perspective of a true developmental science, a multidisciplinary concept that is a worthy successor to the all-too-often static perspective of child psychology.

No book is written alone, and this one is no exception. I want to thank my colleagues in the Human Development and Family Studies Program at the University of Vermont for their support, especially in letting me use some office space for a few months following my retirement. I want to thank Professor Jonathan Tudge for his wise comments in exchanges we have had about nature–nurture and developmental theory in general, even if we continue to disagree on a point or two. I want to thank the reviewers of the prospectus for this book as well as Cambridge University Press for seeing value in my efforts, the editors and copy editors at the Press for doing a wonderful job in helping make this book read as well as it possibly could, and, especially, a very special group of students who read early drafts of the text as part of my last seminar as a member of the faculty; appropriately enough, the seminar was called The Nature–Nurture Debates.

1

Issues and Questions

It's really incredible when you think about it. Here we are, well into the twenty-first century, and we are still fighting over the role of nature and nurture in human development. And it isn't even a new fight; it's not even a twentieth-century fight. It actually goes back to the nineteenth century and probably even before that. So why is it that we cannot get this question answered and move on to a new one? Is it because we haven't yet gotten the necessary data to make a conclusion one way or the other? Do we not yet have a powerful enough computer to sort everything out? Have we not identified the best method and statistics to collect and analyze the relevant data? One answer to these questions is, of course, "yes" to all these possibilities, but there is also another possibility. It may also be that we are having trouble coming up with the answer because we continue to ask the wrong questions.

The Nature–Nurture Debates: Bridging the Gap is an attempt to make sense out of the nature–nurture debate, to explain why this debate is still even a debate. I mean, after all, how many other topics in any of the sciences have been debated for more than 150 years without any resolution? Making sense out of the debate requires an examination of several issues and questions. For starters, what in fact are we talking about when we talk about nature and nurture? How is each measured, and how is its relative contribution assessed? What is the history of the debate? Have there been solutions that we now no longer accept? What were they? Why were they rejected? What has changed in our understanding of the course of human development? How has this change redefined the debate? What difference does it really make anyway how much nature and nurture influence our development? Is this essentially an academic debate that may never be answered to everyone's satisfaction, or are there important practical implications as well? What are the major theoretical positions in the debate? What does each have to offer? What claims does each make? What data does each provide?

Maybe a good place to start answering some of these questions is to agree on some terms. There is no common agreement on how to best describe those who see nature as having a disproportionate influence on development and those who see nurture as being more dominant. One set of terms that does have some historic precedence (Carmichael 1926) and is being used increasingly in the literature (Spelke 1998; Simpson et al. 2005; Spelke and Kinzler 2009) is that of nativists and empiricists. Carmichael (1926), in discussing the meaning of "empirical psychology," notes that the term empirical can be seen as referring to a perspective that sees development as an acquired process rather than an innate one and, as such, "the term is antithetical to nativism" (p. 522).

Having now agreed that we call those favoring nature nativists and those favoring nurture empiricists, the next issue is to get a better definition of what each is talking about, that is, about what in fact are nature and nurture. At first glance it seems that the nativists have an advantage because it might seem easier to define nature as genetics than nurture as environment. After all, aren't genes so specific and environments so vague? Well, yes and no. The genome mapping project has managed to unravel the DNA code, but the findings were somewhat surprising in at least one way. We do not appear to have enough genes to put us together if in fact genes put us together. The genome project identified about 25,000 distinct genes that seem to be involved in protein synthesis, the particular task genes actually have in our bodies. Even an ear of corn requires more genes in the formation of the proteins involved in its formation. To make matters worse, there is little, if any, evidence of a one-to-one correspondence between specific genes and specific outcomes at any biological or behavioral level. Rather, individual genes work in combination with other genes to produce proteins and the same genes, in combinations with other genes, produce different proteins. We even share many of the same genes found in other species. For example, we share approximately 98% of our genes with our closest primate evolutionary relatives and yet we are distinct from them in so many ways. Some researchers believe that the issue then is not what particular genes are present but how they function and how they are regulated. Kagan (2001), for example, notes that the same genes are involved in the formation of our brains as those of chimpanzees but that, in our case, these genes stay active longer in humans, allowing for the additional layers of cortex that apparently are reflected in our significant cognitive advantages. Then there is the matter of the fact that the majority of genes do not appear to code for anything. That is, the majority of our genes have been seen simply as "junk," a vestige of our evolutionary heritage and with no obvious role in our development. However, increasingly we are coming to appreciate (Keller 2010) that much of this junk is not junk at all but rather serves the purpose of regulating the activity of structural genes. We examine the role of genes in more detail in later chapters, but for now, the answers to the question of what is a gene is are (a) it is a lot of different things, (b) we are not as sure as we once thought we were, and (c) we need to understand much more about how some genes go about regulating the activity of other genes, in particular, about what factors regulate these regulatory processes. All three appear to be true.

How then can we define environment? The first problem is the same as for defining genetics, that is, everything is environmental just as everything is genetic. If we define environment as context, then we quickly come to the realization that environment exists at a variety of levels because we can as easily talk of the context in which a protein functions as we can that of a 3-year old. In fact, because we can talk about an individual as existing at a variety of levels simultaneously, from the level of the gene to the level of the culture and beyond, then there are always many different environments impinging on the individual at any one time. To make matters worse yet, at least for humans, we can make a distinction between the actual environment and the perceived environment. You need to spend only a few minutes in a classroom to recognize that even though all the children are experiencing virtually the identical actual environment, the behavior of the teacher, the varied reactions by the children to the teacher's efforts tell us that their individual responses may not so much be to the teacher's behaviors as to their perception of the meaning of that behavior.

Even though we can talk about the environment as existing at a number of different levels across a number of different domains, the nurture side of the debate typically involves the interpersonal environment of interest to psychologists and others interested in the development of children and adults. In other words, the role of nurture in the debate is much more likely to involve some issue related to the family, the education of the child, or the peer group rather than the child's prenatal environment or the impact of pollutants in the air. In effect, one could argue that one reason the empiricists have lately found themselves somewhat on the defensive may be because they typically consider only a limited swath of the full range of the environment, that is, they may be engaging in the debate having tied one hand behind their backs (Shonkoff 2010).

The difficulty of defining both genetics and environment is clearly reflected in the form of the nature–nurture debate. It is much easier to control the behavior of the teacher than the children's perception of her efforts, even though those perceptions may be more developmentally significant. At best, we can only hypothesize about and test for some degree of correspondence between the two. Similarly, if in the biological sense there can be no environmental effect without an organism to act on and if there can be no organism outside of its context, then nature and nurture would appear to be inseparable. But we have statistical procedures that are intended to do just that: partition genetic and environmental influences. How can we reconcile this seeming contradiction? Can it be reconciled?

Part of the problem in attempting to answer all these questions is that in reality, when we talk about the nature–nurture debate, we are not talking about one debate but rather about three debates. The first debate is the seemingly classic one: What are the relative and independent contributions of genetics and environment to an individual's development? This *classic debate* is a reductionist debate and reflects the theoretical view that complex phenomena can be reduced to individual antecedents and the influence of each antecedent can be assessed independently of all others. The second, or *new debate*, is taking place at a more holistic, systemic level. Here both sides seem to recognize that both nature and nurture are essential and ultimately inseparable, but there nevertheless remains in the new debate very much of a "chicken-and-egg" argument. There is some overlap in the participants of the classic and new debates, especially in terms of some of those who would describe themselves as evolutionary psychologists, but for the most part, the two debates are distinguishable by the level at which each is fought. Even though, for example, nativists and empiricists are on opposite sides in the classic debate, they often become strange bedfellows with respect to the new debate by both favoring a reductionist position in opposition to those favoring a more holistic systems perspective.

The third debate is actually a *proxy debate*. In this third case, the debate is simply a meeting place for arguments about even more basic issues, ones that define the very subject matter of human development. The three debates are not truly independent of each other, but it is worthwhile nonetheless to discuss the three separately because each sheds some light on why the debate (debate*s*, actually) never seems to move off center, much less come to some resolution. It probably makes the most sense to start with the third debate because it raises the most basic questions about human development.

The Proxy Debate

Open up just about any introductory child development or life-span development text, and in the first chapter, in addition to the usual homage to Freud, Piaget, Erikson, Skinner, and the other grand theorists, will be a discussion about method and theory. The method discussion will talk about such things as independent and dependent variables, the difference between correlational research strategies and controlled research strategies and about statistical analyses that allow you to measure the unique influence of each independent variable on the dependent variable. The theory section will most likely discuss whether development is best conceptualized as a continuous or discontinuous phenomenon, as showing individuals' status as relatively stable or changing compared with others over extended periods of time, the relative importance of structure and function in influencing behavior and behavior change, and the degree to which behavior primarily reflects preprogrammed or innate variables as opposed to environmental or epigenetic variables. The particular choice of words to describe each of these topics might differ, especially with respect to the "predefined" aspect of our development, but all these texts will have essentially the same discussion. And then the first chapter comes to an end, and, depending on whether the text is organized chronologically or topically, the next chapter concerns either prenatal development or perhaps biological development, and so on. Interestingly enough, the methodological and theoretical issues raised in the introductory chapters rarely, if ever, reappear in subsequent chapters. The reader is left with the impression that either these first chapter issues have in fact been resolved long ago or maybe they really are not that relevant to understanding the details of a particular developmental stage or domain.

The fact of the matter is that even though these methodological and theoretical issues might seem resolved to the reader of that introductory text, they are anything but, and more often than not it is through the nature–nurture debate that this proxy debate continues. Cronbach (1957) noted this rift many years ago when he talked of the distinction between what he referred to as "experimental psychology" and "correlational psychology." Whereas experimentalists are interested in only the variability that they are able to create through variations in experimental conditions, correlationalists are interested in examining the already present variabilities among individuals, groups, and species. Cronbach (1957) says that, for the experimentalist, individual differences are "an annoyance" because they reflect the "outer darkness known as error variance" (p. 674). But the correlational psychologist is "in love with those variables the experimenter left home to forget" (p. 674). For the correlational psychologist, the question of interest is how the already present characteristics of individuals determine their mode and degree of adaptation. And although we do not often see Cronbach's terms in use today, given the respective interests and typical methodologies of nativists and empiricists, we are still seeing the same two disciplines that he noted. Nativists, in the classic debate, are primarily interested in individual differences. They continue the correlational tradition, although their methods are not necessarily restricted to correlational statistics. Empiricists, in the classic debate, are the experimentalists. They are interested in the identification and control of those variables that regulated development patterns and change.

McCall (1981) did not find things much improved 25 years later. He also noted a distinction between those developmentalists primarily interested in developmental functions common to all members of a species and those interested in the relative consistency of individual differences among members of the same species over time, a distinction he equates with empiricists and nativists, respectively. He saw this difference as reflecting the two realms of development and makes the point that continuing this gap hinders our full understanding of development. In particular, factors that may influence individual differences may have little influence on developmental functions and vice versa. For example, the variables that influence species-typical behaviors, such as walking or other large motor skills, may have little, if anything, to do with the factors that influence variability in onset or competence in walking or skipping or throwing. McCall notes that a focus solely on individual differences is like "studying the consistency of a few inches in the heights of giant sequoia trees from seedlings to maturity while ignoring the issue of how all the trees grow to over 300 feet" (1981, p. 3). He is not more sympathetic to empiricists:

> Lest the environmentalists feel smug, they are no better off. The environmental as well as the genetic factors necessary to produce fundamental characteristics in the species are available to almost everyone we study. As a result, the only way we can study the importance of certain major environmental factor for development is to take advantage of tragedies – children reared in closets, born blind and later given sight, or fed from birth through a fistula. (p. 4)

What Cronbach referred to as the two disciplines and McCall referred to as the two realms are no closer today; if anything, they are perhaps farther apart because of our presumed greater ability to study genetic and biological processes. The distinction is as evident in the distinction Simpson et al. (2005) make between nativists and empiricists:

> Nativists are inclined to see the mind as a product of a relatively large number of innately specified, relatively complex, domain-specific structures and processes. Their empiricist counterparts incline toward the view that much less of the content of the mind exists prior to worldly experience, and that the processes that operate upon this experience are of a much more domain-general nature. (p. 5)

In other words, nativists are more likely to see development as a continuous expression of some number of predefined capabilities, traits, or modules, each exerting its influence relatively independently of the others, whereas empiricists are much more likely to see development as an epigenetic, probabilistic process, one much more dependent on the vagaries of the lived experience. For nativists, the focus then becomes on understanding how these predefined variables cause differences between individuals and the degree to which these individual differences are stable over time. For empiricists, since little is seen as inherent, the focus is on an examination of the role external variables play in structuring the life course and the degree to which changes in these antecedents are predictive of changes in an individual's behavior. It is not that nativists deny any role for the environment or that empiricists deny any role for biological structure; rather it is that each see that other in, at best, a supporting role.

Said another way, the proxy debate is not restricted to arguing the relative merits of nature and nurture so much as it is a debate about the very foundation and maybe even soul of the discipline of human development. And it is a debate that goes deeper than issues of stability versus change. It is a debate that also argues the legitimacy of types of causes, the place of the concept of "purpose" in the study of human development, and even what the "original" causes of our development are.

Nativists tend to place most emphasis on what are seen as material causes and empiricists place that emphasis on efficient causes (Pepper 1961; Goldhaber 2000). Material causes are seen as components of the individual, such as the presumed modular structure of the brain or ones particular genotype. Efficient causes are seen as factors external to the individual; they are things that happen to the individual. Embedded within the classic debate at least is the fundamental belief that, although both efficient and material causes influence development, the two can be disentangled from each other and the relative influence of each determined independently of the influence of the other. Discovering such origins is seen as the fundamental purpose of science.

Even debates about the place of teleology in the study of development are reflected in the proxy debate. Teleology is the belief that ends are immanent in nature and that natural phenomena are determined not only by mechanical causes but an overall design as well (Anandalakshmy and Grinder 1970). Such notions are often reflected in evolutionary arguments about development that see the *purpose* of behavior to be increasing the chances that the organism will reproduce and therefore that the species will survive. Empiricists see little, if any, value in teleological arguments, placing them more in the realm of philosophy and religion than in that of science. For empiricists, development is largely a reflection of the particulars of time and place.

The Classic Debate

The classic debate is between nativists and empiricists who look to the environment as the primary determinant of development. The point of the debate is the relative contributions of genetic and environmental influences on the course of human development. It is predicated on the shared (by both nativists and empiricists) fundamental belief that antecedents can be partitioned and the relative influence of each ascertained. This is where the agreement ends and the debate begins because each side favors a different set of antecedents, often examined with different methodologies, in one case measures looking for stability and in the other measures designed to look for change.

The nature side is represented by those who would describe themselves as *developmental behavior geneticists.* The nurture side, however, is a little tricky to define. There really isn't any particularly identifiable group so much as there are a large number of developmentalists examining the influence of any number of external antecedents on behavior. In fact, it is this lack of a definable environmental perspective, coupled with the significant advances in our understanding of genetic and biological processes over the last several decades, that has made it possible for the nature side to become increasingly visible and influential both within the discipline of human development and more broadly across the culture. Nativists take on the empiricists all the time, but the reverse is rare. More often than not, empiricists do not initiate the debate so much as they respond to nativists claims, as was the case when Jensen (1969) published his famous (or infamous, depending on your view) *Harvard Educational Review* article arguing that preschool test score gains in response to educational interventions were little more than a "hothouse" effect or when Scarr (1992), in her 1991 presidential address to the Society for Research in Child Development, made her argument for the "good enough parent." Both publications quickly brought forth rebuttals, but otherwise, empiricists do not seem to feel the need to challenge nativists in the same way that nativists challenge empiricists.

The particulars of the classic debate are discussed at length later in the book, but for now a sampling should make clear why this debate has gone on for so long and will continue to go on forever as it is presently constituted. The sampling concerns Scarr's reference to good enough parents and the response of one of her rebutters.

Scarr, speaking to a group whose members then and now are predominantly empiricist in orientation, made a striking behavioral genetic argument as to the relative influence of genes and environment on children's development. Simply put, according to Scarr, genotypes drive experience. Parental genes determine parental phenotype, child genes determine child phenotype, and the child's environment is "merely a reflection of the characteristics of both parent and child. Here differences among children's common home environments, *within the normal species range* [italics in original] have no effect on differences among children's outcomes" (1992, p. 9). And to support her argument, Scarr reported several kinship and adoptions studies that found much higher correlation coefficients among those more closely related (e.g., identical twins compared with fraternal twins) and, interestingly enough, between children and their biological mothers (even when separated at birth) than their adoptive mothers.

Scarr saw her strong genetic message as comforting to parents because she saw it as relieving parents of the burden of trying to be perfect. Now they just need to be "good enough" to keep their child's development within a typical, expected trajectory. And if they are good enough and if the child still goes off on an "undesirable trajectory," interventions are likely to have only limited and temporary effects.

Well, you can imagine the reaction from many in the audience on hearing that individuals make their own environments, based on their heritable characteristics. I do not know if Diana Baumrind was one of those actually in the audience that day, but the following year, she published (Baumrind 1993) a rebuttal to Scarr's address titled "The average expectable environment is not good enough: A response to Scarr." Her rebuttal took three forms. First, she questioned Scarr's conceptualization of a good enough environment, claiming that the concept was ill-defined and therefore of little scientific value. Second, she raised several methodological issues with the data Scarr reported, both in terms of how it was collected and the statistical procedures used in their analyses. Third, she reported a number of studies, her own work included, that she saw as clearly showing a significant parental influence on child outcome. In her words,

> There is a large and consensually validated body of evidence relating to children's prosocial competence to such parenting skills as persuasive communication, contingent reinforcement, and monitoring; and children's cognitive development to such parenting skills as scaffolding, academic engagement, and high-level distancing skills. All these parental practices manifest a high, not good enough, level of parental involvement and commitment. (pp. 1311–12)

The relative merits of Scarr's and Baumrind's positions aside, the one thing that is clear from this typical type of exchange between nativists and empiricists is the fact that they simply cannot agree on the terms of the argument beyond a commitment to reductionism. And if you cannot agree on the ground rules, then you simply cannot resolve the issue.

The New Debate

The new debate is between those who approach the study of development from an *evolutionary psychological perspective* and those who approach the study of development from a *developmental systems perspective*. It would be incorrect to equate one group with nativism and the other with empiricism as most in each camp recognize the synergistic interplay between nature and nurture.

Evolutionary psychology considers how our evolution as a species has come to influence our current behavior and development (Pinker 2002; Tooby et al. 2005; Geary 2006). We, as a distinct species, emerged approximately 100,000 years ago during the Pleistocene era. The characteristics that came to define us at that time reflected the cumulative consequences of adaptations to those conditions existing during the few millions years of hominid evolution. Evolutionary psychology argues that because we are, genetically at least, essentially the same as we were 100,000 years ago, the same genetic mechanisms that regulated our behavior and development then do so as well today. Needless to say, this is not a perspective that has gone unchallenged either in terms of the behavioral implications of such an argument as to the accuracy of the claims about our genetic similarity to our Pleistocene ancestors or in terms of arguments relating to the degree of flexibility of our genome (Ehrlich and Feldman 2003).

This evolutionary perspective has served also to reevaluate the concept and place of innateness in the developmental process. Have we, as a species, as a result of our evolutionary history, evolved certain specific structural domains or modules (Spelke and Kinzler 2009) that regulate our behavior to some measurable degree? Are these structures present at birth? How specific are they? How flexible are they? The renewed interest in the competencies of very young infants is in part a reflection of this interest in evolutionary psychology's arguments for a high degree of domain specificity present at birth.

Developmental system theory (Lewis 2000; Johnston and Edwards 2002; Gottlieb 2003; Lickliter 2009), on the other hand, argues that all developmental forms emerge out of the recursive bidirectional interactions of simpler components. The properties that emerge from these interactions are self-organizing and are not reducible in origin to any of the individual contributing components. So, for example, the elements in the developmental sequence of a child's motor development would be understood from this perspective as the emerging results of the continuous interplay between the child's existing biological structures in interaction with the child's evolving environmental context (Thelen 1995). References to existing biological structures from a developmental systems perspective are not meant to imply activity of some sort of genetic regulatory mechanism but rather the cumulative interaction history of the child up to that point in his or her development.

Whereas the classic debate is fought at a reductionist level, the new debate is being waged at a more systemic, holistic level, at least for most participants. This caveat reflects the fact that some evolutionary psychologists seem to prefer reductionist explanations for the expression of behavior whereas others are more comfortable at a systems level. Whereas the classis debate will never be resolved, the new one does have a glimmer of hope.

The distinction between the classic and the new debates is as much a matter of a debate over the appropriate level of analysis to study development as it is a debate over the relative roles of heredity and environment. As previously discussed, the antagonists in the classic debate each believe that complex phenomena can be reduced to a set of independent individual causes and the distinct influence of each of these causes can then be determined. Those involved in the new debate believe that this reductionist model is simply wrong. They argue that you cannot reduce complex phenomena to independent constituent components. Or, more precisely, that in attempting to do such partitioning through statistical analyses, you lose the very phenomena you are interested in better understanding. Said another way, one can certainly study the properties of oxygen independently of those of hydrogen, but the study of either will not provide any insight into the properties of water. It is through the interaction of hydrogen and oxygen that the properties of water emerge. So too is the case with those working at a systems level with respect to human development. Consider the contrasting views of first Meany (2001) and then Pinker (2004) on this issue. Meany approaches the issues from a systems perspective whereas Pinker chooses a reductionist perspective.

> So too it is with "nature" and "nurture," for life does not emerge as a function of either. It is equally wrong headed to assume that, oh yes, phenotype derives from both nature and nurture. This would only be to repeat the misunderstanding in kinder, gentler terms, as if biological and social scientists had shaken hands and then gone off into their own corners of the universe to study "lengths" and "widths." Indeed, both conclusions derive from additive models of determinism

> where gene + environment = phenotype. Such models make no biological sense whatsoever. It is not nature *or* nurture. Nor is it nature *and* nurture.... There are no genetic factors that can be studied independently of the environment, and there are no environmental factors that function independently of the genome. Phenotypes emerge only from the interaction of gene and environment. The search for main effects is a fool's errand. (Meany 2001, p. 51)

> But the very thing that makes holistic interactionism so appealing should also make us wary of it. No matter how complex an interaction is, it can be understood only by identifying the components and how they interact. Holistic interactionism can stand in the way of such understanding by dismissing any attempt to disentangle heredity and environment as uncouth. (Pinker 2004, p. 8)

So what for Meany is the only level of analysis, a systems level, that allows for an understanding of development is to Pinker the very impediment standing in our way to being able to understand human development. Is one right and the other wrong? Clearly each is talking about a very different level of analysis. Therefore, the question really becomes one of asking if one level is the correct one and the other the incorrect one or is it possible that some developmental phenomena may be best understood at a reductionist level and some best understood at a systems level? In other words, even if nature and nurture interact in such a way that all developmental phenomena are ultimately a reflection of the synergistic action of both, is it still possible for some phenomena, at least in a practical sense, to consider data obtained through reductionist research strategies valid?

Plan Of The Book

The remaining chapters of this book attempt to provide some answers to all of these questions. This first chapter has laid out the basic issues and has identified the "three debates." Chapter 2 takes a historical perspective on the debates. It makes clear the origins of the debate, especially the classic debate, and the early attempts to resolve it. Not surprisingly, these early strategies for resolution appear not that different from the current ones in the classic debate. Looking at the nature–nurture debate from a historical perspective provides some insight into how we have gotten to the current point of the debate. Chapter 3 examines the proxy debate. It looks at the issues of metatheory, methodology, and data analysis because much of the proxy debate hinges on the appropriateness of the strategies used to collect and analyze data and on the issue of the most appropriate level of analysis to use in the study of human development. Chapter 4 looks much more closely at the classic debate. It first reviews the developmental behavior genetic position in terms of its basic assumptions, methodology, and data. The chapter then looks at the environmental or empiricist side of the classic debate although, as I noted earlier, this is not as easy as it might seem because there really is no environmental position but rather a lot of environmental positions, each relevant to a particular area of study. Chapter 5 examines the new debate by reviewing both the evolutionary psychological position and the developmental systems perspective. It is not so much as these two actively oppose each other as it is that one takes a very distal perspective and the other a proximate perspective. The fundamental issue in the new debate is not which is right but rather how possible it is to integrate these two very distinct time perspectives. Chapter 6 asks the question, "So What?" Does it really matter how the debate is structured or resolved? The answer clearly is that it does because, unlike other topics within the study of human development, interpretations of the role of nature and nurture have touched and continue to touch virtually every aspect of our lives. It is reflected in our immigration policies, our educational strategies, civil rights legislation, matters of gender equity, and strategies to end or, in the case of some, simply recognize the inevitability of social inequality, to name just a few. Finally, Chapter 7 offers some modest suggestions on where we go from here. And without giving the whole plot away in the beginning of the story, suffice it to say that our only hope in resolving the nature–nurture debate rests with the new debate because it is the only perspective that offers a contemporary perspective on the interplay of nature and nurture and therefore the only one that offers a realistic portrayal of development across the life span.

2

A Brief History Lesson

One of the things that all students of human development are taught in their research methodology and statistics classes is that methodology and statistics are designed to be neutral, that is, the results of an experiment should reflect the data collected rather than how the data were collected or analyzed. This is the bedrock, the basis on which we claim our work to be objective, to be scientific. However, at least with respect to nature and nurture, and, in particular, the early history of the classic debate, this fundamental assumption simply doesn't appear to be true. In fact, most of the research methods and statistics used in the debate, then and now, evolved out of a desire of those whom we would now recognize as nativists to have an "objective" way of demonstrating the dominant position of nature over nurture rather than an objective means to assess the relative importance of the two.

Like just about everything in Western thought, this brief history lesson goes all the way back to the ancient Greek philosophers, Plato (437–347 B.C.E.) and Aristotle (384–322 B.C.E.). Today we would call Plato a nativist (Simpson et al. 2005), although it was not a term he made use of. Nevertheless, Plato argued that experience was simply insufficient to account for all the knowledge and abilities humans possess. Because these things cannot be taught, they must instead be present at birth, that is, they are innate. On the other hand, Aristotle, Plato's student, might have been the first empiricist, more precisely, possibly the first epigeneticist. As Anandalakshmy and Grinder (1970) explain it,

> Aristotle described development as a process of continual integration and differentiation and insisted that the more complex growth patterns were irreducible to simpler elements or atoms. From the epigenetic viewpoint, maturity is not simply a summation of earlier structures but a novel synthesis of them. (pp. 1118–19)

Neither Plato nor Aristotle apparently actually used the terms "nature" and "nurture"; that honor goes to Richard Mulcaster in the year 1582 (West and

King 1987), who spoke of the harmonious relationship between the two, how nature and nurture come to collaborate in defining a child's development. Mulcaster would probably be rolling over in his grave today if he knew how unharmonious the debate over nature and nurture has turned out to be.

These early positions on nature and nurture were philosophical rather than empirical. They reflected the beliefs and the logical reasoning of the individual rather than any sort of data, and they reflected the general sociohistorical context in which such thinking was taking place. Such thought in context was equally evident in the philosophies of both John Locke (1632–1734) and Jean-Jacque Rousseau (1712–78) (Gianoutsos 2006). Locke's philosophy is often equated with seeing the child as a "blank slate" onto which knowledge and morals are written. Although his views aren't quite so absolutely environmental, they nevertheless reflect a role for experience clearly consistent with a bias toward nurture. Rousseau, on the other hand, saw children coming into the world as endowed with goodness. The proper role for the environment then is to leave the child alone and the proper role for parents is to shield the child from the evils of experience. Here too, this is not as extreme a nature view as sometimes ascribed to Rousseau, but nevertheless it does seem to favor the role of nature over nurture.

The philosophies of these two men are reflected in debates about child rearing, about education, and even about the proper role of government. Pastore (1949), for example, sees Locke's belief about the power of society as a positive force providing one of the major conceptual underpinnings of both the American and French revolutions. He also notes that critics of "liberty, equality, and fraternity" saw such emancipatory efforts as radical and serving only to upset the "natural order" of what is good and right and replace it instead with a tyranny of the masses (which of course others might now refer to as democracy).

So in many ways the stage was set for the emergence of the classic debate: Enter Charles Darwin (1809–82) and his cousin Francis Galton (1822–1911). It is impossible to overstate the impact of Darwin's theory of evolution on our understanding of development. Just as Copernicus helped us understand our place in the universe, Darwin's theory helped most of us at least to fully appreciate our place with respect to all other species. As profound as Darwin's ideas were, however, he actually had little to say specifically about the relative importance of nature and nurture. After all, even if there is a unit of inheritance that passes information across generations, the ecological niche of a species ultimately has as much to do with the actual passage of that message and, for that matter, if there will even continue to be future generations of that species (West and King 1987). It was rather Darwin's cousin Francis Galton who saw in a theory of evolution a way to differentiate nature from nurture and then to ascribe what for him was the rightful importance of each. It is really Galton who was the first to see the roles of nature and nurture as distinguishable and perhaps of greater importance to the debate, as existing as oppositional forces, each competing to influence development. For Galton, nature was clearly the winner.

Galton interpreted Darwin's theory to mean that the process of evolution could be seen as a ladder, with different species occupying different rungs on the ladder. Those higher up were more advanced than those below. This is not, in fact, a very accurate picture of evolution now or then (Gould 1996) because the relationships between species are better represented through a tree than a ladder. Further, because the viability of each species is measured by its survival, it is hard to argue that we as a species are more advanced than even the dinosaurs because they managed to stick around a lot longer than we have to date. But to Galton, seeing evolution as a ladder of progress made it easy to go the additional step and not just order species but even groups within a species; hence the introduction of "social Darwinism."

The study of development actually owes much to Galton besides social Darwinism. He is credited with being one of the very first to emphasize measurement and experimentation as the best strategy to understand development. Pastore (1949) credits him with introducing questionnaires as a means to gather data, as being the first to recognize the presumed value of the study of twins in unraveling the influence of nature and nurture, and for the development of the correlation coefficient as a statistical tool to measure the degree of association between abilities and degrees of kinship. Fancher (2009) sees his work as laying the conceptual and statistical foundations for the development of tests designed to measure intelligence and for the development of what we now recognize as developmental behavior genetics.

Galton's studies of genius chronicled the accomplishments, across generations, of eminent Englishmen of science, the law, business, the arts, and the military (Galton 1869). Not surprisingly, he found that eminence ran in families, and, as such, he concluded that nature must play a more important role in determining achievement than does nurture. Of course, a more contemporary interpretation of these data would not come to the same conclusion, but what is important to understand about Galton and those who shared his views is that the "data" served to scientifically confirm what he believed to be an obvious truth.

But Galton was not satisfied with simply documenting what he saw as natural differences between individuals within groups and, by extension, differences across groups. His real concerns were the social implications of his data:

> I PROPOSE [all capitals in original] to show in this book that a man's natural abilities are derived by inheritance, under exactly the same limitations as are the form and physical features of the whole organic world. Consequently, as it is easy, notwithstanding those limitations, to obtain by careful selection a permanent breed of dogs or horses gifted with peculiar powers of running, or of doing anything else, so it would be quite practicable to produce a highly-gifted race of men by judicious marriages during several consecutive generations. I shall show that social agencies of an ordinary character, whose influences are little suspected, are at this moment working towards the degradation of human nature, and that others are working towards its improvement. I conclude that each generation has enormous power over the natural gifts of those that follow, and maintain that it is a duty we owe to humanity to investigate the range of that power, and to exercise it in a way that, without being unwise towards ourselves shall be most advantageous to future inhabitants of the earth. (Galton 1869, p. 1)

How fortunate to be able to "conclude" by the very first page of the study. Galton was proposing nothing less than the practice of eugenics. For Galton, the process of evolution provided for the "natural progress" of those more fit to succeed and those less fit to fade away. However, modern civilization was increasingly interfering with this natural process. To Galton, this interference can serve only to make us less fit, to allow those less capable to occupy positions of importance, to "mongrelize the race" through intermarriage, and to lower the overall vitality of the race by allowing those less able to "breed at will." The result of the eugenics movement in England, the United States, and elsewhere was the establishment of laws allowing for the forced sterilization of those deemed less fit and laws against miscegenation. To put these laws in some perspective, it was not until 1967 that the U.S. Supreme Court ruled that such laws that were then still in effect in approximately twenty states, North and South, were unconstitutional, and the last forced sterilization was performed as recently as 1981 in Oregon and in total affected as many as 60,000 Americans (Lombardo 1996).

Galton's perspective was not the exception to the role of nature and nurture. He was very much in the mainstream of his time. Karl Pearson, whose name is immortalized in the minds of all students of statistics who come to learn about *Pearson's r* correlation coefficient, argued that the selection of parentage is the only factor determining the social progress of a race. When, in his words, the dull and idle have no chance to propagate their kind, nations progress even if the land be sterile (Pastore 1949). Similarly, R. A. Fisher (1918), generally recognized as the founder of modern statistics (Mather 1964; Yates 1964; Rao 1992) and, in particular, the developer of factorial research designs and the analysis of variance was equally supportive of eugenic policy. In his own words,

> ... eugenics comes at an appropriate time, when our civilization is already sadly acknowledging that the great bar to progress lies in human imperfection; for the first time it is made possible that humanity may improve as rapidly as its environment. The supposed conflict between heredity and environment is quite superficial; the two are connected by double ties: first that the surest and probably the quickest way to improve environment is to secure a sound stock; and secondly that, for the eugenist, the best environment is that which effects the most rapid racial improvement. (Fisher 1914, p. 310)

In fact, so certain was Fisher of his beliefs that even when his calculation of kinship correlations showed that only 54% of the variance was due to ancestry alone, he concluded that the rest could not be ascribed to the effects of environment but rather to sampling and measurement errors because "it is very unlikely that so much as 5 percent of the total variance is due to causes not heritable" (Fisher 1918, p. 424).

Before we consider the degree to which these ideas were evident in the early work of American psychologists and others, an important caution is warranted. It would be incorrect to assume that those who now use the research designs and statistical techniques initially developed by Galton, Pearson, and Fisher necessarily share these individuals' views on the nature–nurture debate or on its presumed implications for social policy. These designs and methods of analysis are the essential tools used by all developmental researchers, irrespective of how each views the debate. But at the same time it is crucial to remember that these techniques were initially developed for the sole purpose of providing the scientific foundation for the beliefs held by these individuals. The question then becomes whether these techniques are therefore value neutral or whether in some way or to some degree they do in fact bias data in ways consistent with the beliefs of their originators. We will look more closely at the issue in the next chapter, but for now let's move on to the America at the dawn of the twentieth century.

The views of Galton, Pearson, and Fisher enjoyed a warm welcome on this side of the Atlantic in the early part of the twentieth century. However, by the 1920s, a clear empiricist voice was also emerging, and this early edition of the classic debate was in full bloom. The debate, driven both conceptually and empirically, lasted up to World War II. As the atrocities perpetrated by the Nazis on Europeans in the name of eugenics became increasingly evident, public statements by nativists in this country and elsewhere became increasingly scarce. The dominant voice that then emerged in the 1950s and 1960s was a strong empirical behaviorist voice, either reflecting the Hull–Spence (Hull 1943; Spence 1956) tradition or the Skinnerian (Skinner 1938, 1953) tradition. And then, in 1969, with the publication of Arthur Jensen's *Harvard Educational Review* article (Jensen 1969) "How much can we boost IQ and scholastic achievement?," the current edition of the classic debate began. Because all of these elements have had and continue to have an influence on both the classic and new debates, it is worth looking at them a bit more closely.

The early American nativist's work reflected the efforts of Lewis Terman, Henry Goddard, and Robert Yerkes (Hunt 1961; Kamin 1974). Terman taught at Stanford, Goddard was director of the Vineland Training School in New Jersey, an institution providing residential services for those who were then considered to be mentally deficient, and Yerkes taught at Harvard. All three shared a strong nativist perspective, supported eugenic proposals, and had a strong desire to make the new discipline of psychology visible and relevant to society. This nativist perspective led each to focus on issues of individual differences among individuals and groups of individuals, especially with respect to intelligence, and, given their beliefs that these differences were both fixed and genetic in origin, a logical expression of their work was in the area of mental testing, in particular, the Binet IQ test.

Alfred Binet developed his test in 1905 as a way to identify those French children who might be in need of specialized educational services. There was nothing in his work that implied a theoretical bias; it was simply seen by him as a pragmatic diagnostic instrument. Terman, Goddard, and Yerkes were instrumental in translating the Binet test and bringing it to the United States. But they did more than simply translate the test; they redefined a "pragmatic diagnostic instrument" into an instrument that "could be used to provide statistical support for the already demonstrated proposition that normal intelligence, and 'weak mindedness' were the products of Mendelian inheritance" (Kamin 1974, p. 7). The Binet and other instruments derived from the Binet were then used to test recruits in World War I as well as the thousands of immigrants coming to this country, primarily from Eastern and Southern Europe. The results of these testing efforts were certainly a serious concern to these three as well as to others sharing their views. Goddard, for example, at the request of the U.S. Public Health Service administered the Binet test to arrivals at Ellis Island and, based on his analysis of the test data, concluded that 83% of the Jews, 80% of the Hungarians, 79% of the Italians, and 87% of the Russians were feeble minded (Kamin 1974). None today would accept such data, but in 1912, when Goddard reported his findings, people in positions of power did take notice. In fact, they did more than take notice: They introduced in Congress legislation restricting the immigration of people from Eastern and Southern Europe, in effect, establishing national origin quotas:

> The hereditarian theory of IQ is a home-grown American product. If this claim seems paradoxical for a land with egalitarian traditions, remember also the jingoistic nationalism of World War I, the fear of established old Americans facing a tide of cheap (and sometimes politically radical) labor emigrating from southern and eastern Europe, and above all our persistent, indigenous racism. (Gould 1981, pp. 157–8)

But as powerful as these nativists voices were, they were not the only voices trying to define either the discipline of psychology or the public agenda. By the 1920s, an increasingly strong empiricist voice was also emerging. Edward Thorndike is probably a good person to start with, as he seemed to have one foot firmly in the nativist camp and the other equally firmly in the empiricist camp. His beliefs about the role of heredity were very much in agreement with people such as Terman and Goddard. He too expressed concern about the "masses," favoring instead a form of intellectual and moral "aristocracy" in which those "most able" would make decisions most consistent with the preservation of the species, that is, eugenics. He was even very pessimistic about the role of education to improve humankind, arguing that

> If one has imagined giving the intellectually underprivileged the advantages of a home where the parents have able minds and encourage intellect in their offspring would cause the genes of a moron to develop into a mind equally to that of the average present-day European, or cause the genes of a 'dull-normal' to develop into a mind able to graduate from a reputable law school, he will be disappointed to learn that differences in home life and training probably cause less than a fifth of the variation among individuals in I.Q. (Thorndike, quoted by Pastore 1949, p. 70)

Clearly this is a sentiment consistent with the nativist thought of the time, nothing remarkable about it except for the fact that Thorndike is generally seen as the father of educational psychology and one of the earliest proponents of the laws of learning. Thorndike's empirical practice, as opposed to his nativist beliefs, focused on animal learning, and it is from this work that the basic laws of learning were first made explicit. His *law of effect,* stating that the recurrence of a behavior is primarily a function of the consequences of that behavior on earlier occurrences is, for example, the basis for Skinner's later work on operant conditioning. Equally influential was his *law of recency,* a second condition influencing the strength of an association between an action and the consequences of that action. Apparently his animal research, clearly demonstrating a strong empiricist perspective, never transferred to his understanding of individual differences in humans, a strange quirk given that one of Thorndike's other areas of scholarship dealt with "transfer of training effects."

John Watson, on the other hand, had both feet pretty much planted in the empiricist camp. Watson is generally credited with being the founder of modern behaviorism and of demonstrating the importance of both classical and instrumental conditioning (Horowitz 1992; Goldhaber 2000). His "Little Albert" experiment, as unethical as we now recognize it to be, is well known to most introductory psych students and involves the demonstration of creating a classically conditioned response by means of pairing an unconditioned stimulus, in this case the noxious sound of striking a steel bar, with a conditioned stimulus, a white mouse that 11-month-old Albert was fond of. Sure enough, after a few pairings, the sight of the white mouse alone elicited the fear response that previously only the sound of the steel bar had elicited. For Watson, it was the environment that created the conditions of one's status; there simply was no need to talk about anything inside of the organism when you could demonstrate an association between an external event and a subsequent behavior. He certainly acknowledged that he often went beyond his data (Watson 1930) when he would make claims about being able to direct people into whatever future he might envision, doctor, lawyer, beggarman, or thief, but he was nevertheless offering, in contrast to men like Terman, what came to be seen as an optimistic, hopeful, egalitarian view of human development (Stevenson 1983).

Although certainly less well known today than those of Watson, it is nevertheless important to recognize the views of Zing-Yang Kuo (1924, 1929) in the early rounds of the classic debate, both because he also took a strong empiricist view and because his later work (Kuo 1976) with Gilbert Gottlieb set the stage for what has become one side of the new debate, the developmental systems side. As was true of most psychologists of those early years, Kuo made a clear and sharp distinction between biology and psychology. No doubt this was seen as necessary for the new discipline of psychology to acquire legitimacy on its own merits, but in so doing, these early psychologists virtually put in place what we now recognize as an arbitrary distinction that has almost certainly done little to advance our understanding of human development and has allowed the classic debate to go on and on and on. Kuo's words make this distinction very specific:

Heredity is concerned with definite physio-morphological problems. It is fundamentally a biological problem. But it is negligible in the study of behavior. We can accept the organism as given and start to investigate its behavior in response to environmental stimulation without reference to heredity. We need not ask how the organism comes about, or how heredity determines the organismic pattern. This is a question the biologist must answer. When the biologist delivers us an organism of a given species in a given stage of development, our duty is to find out how and what stimuli can effectively force this organism to behave and in what manner it behaves. Behavior is not a manifestation of hereditary factors, nor can it be expressed in terms of heredity; it is the direct result of environmental stimulation. (1929, pp. 196–7)

So, clearly, in the 1920s, the classic debate was in full flower, making virtually the identical arguments about the role of heredity and environment as are made today by the current generation still slugging it out over which is more important. What about the "new debate"? Is there any evidence that the new debate is also beginning to take place? "A little" is the answer. Certainly Kuo was beginning to move in that direction through his research on factors influencing the development of chick embryos, and there was one paper that did make a very strong case for what we now would recognize as a systems level resolution of the debate. This was an article by Leonard Carmichael (1925) titled, appropriately enough, "Heredity and environment: Are they antithetical?"

Carmichael was an early proponent of a psychobiological approach to the study of development, and much of his work focused on variables influencing prenatal development. You find little reference to his work today (Logan and Johnston 2007), and this may reflect the fact that Carmichael left his academic work early to take up a series of senior administrative posts at Tufts University, the Smithsonian Institute, and eventually the National Geographic Society. Nevertheless, what he had to say then is as important to consider today as it was in the 1920s.

Carmichael's article was a critique of the nativist's position, in particular of the argument that was common then (and now) that each species has a set of instincts or capacities that are largely defined by genetics and that are little influenced by environment. He chastised nativists for giving more importance to genetics than even the geneticists at the time were willing to give. Carmichael argued that the very fact that nativists could not agree then (or now) on even the number or nature of these inborn capacities made such arguments specious. Further, such nativist arguments denied the possibility that particular behavior patterns present at birth or emerging later might as likely reflect prenatal interactions as solely genetic influences. He noted that small changes in the prenatal environment can lead to significant structural and behavioral changes in species such as fish or amphibians. Arguing that such abnormal changes cannot be "preformed," Carmichael (1925, p. 250) concluded that "These changes develop as a result of the interplay of heredity and environmental factors, and so in order to have a 'normal' individual it is necessary to provide a very specific environment in which development is to take place." The strong psychobiological argument, which we would now describe as epigenetic, was attempting to both broaden the definition of psychology to include any and all forms of stimulation, not just ones the organism might encounter postnatally, and to position psychology and biology as allies in the study of development rather than as antagonists. The debate might well be very different today if his conclusion in the 1925 paper had received greater consideration:

> In man, from the first environmental stimulation of the fertilized ovum until, it may be, well past three score and ten years, the human individual is not made up of two substances: one acquired, the other innate. The human organism and personality, rather, is a unity produced by both of these forces. The unique resulting totality cannot profitably be violated by a destroying analysis, and dichotomized as part native and part acquired.
>
> If this view is true, and we have tried to indicate the nature of the evidence which makes it seem an unescapable [sic] conclusion; the question of how to separate the native from the acquired in the responses of man does not seem likely to be answered because the question itself is unintelligible. Like the old attempt to separate form from material, the effort to sever the modifications due to the environment from those which are innately given, is impossible, save at the level of sterile, verbal abstraction. (p. 258)

In spite of Carmichael's chastisement, the classic debate nevertheless continued in earnest in the 1930s, but its character changed. Whereas the form of the debate in the 1920s was in large part rhetorical, by the 1930s it became much more data driven. In fact, the standard research paradigms that define the classic debate's battleground today all became well established during the 1930s. These paradigms fall into four categories: (a) comparisons of individuals differing in degree of genetic relatedness, (b) studies comparing biological and adopted children with their biological and adopted parents, (c) studies of environmental interventions in the case of humans, and (d) studies of environmental deprivation in the case of other species. The logic of these research paradigms makes perfect sense if one functions within the assumptions of the classic debate, that is, the ability to isolate variables and attribute a portion of variance to each independently of the others. To those who do not share the perspective of the classic debate, then these paradigms, in Carmichael's (1925) words, amount to sterile, verbal abstraction. The next chapter looks more carefully at issues of method and analysis, but a short explanation of each of these paradigms is appropriate here.

If heredity and environment are independent factors influencing development, then adoption and kinship studies are seemingly perfect vehicles to determine the relative influence of each, at least as far as the nativists taking part in the classical debate were concerned. So, for example, a study comparing the similarity of monozygotic (MZ) and dizygotic (DZ) twins would presumably provide a clear measure of the respective roles of heredity and environment because genetics is, in effect, being controlled for. The logic is simple, if ultimately controversial. It rests on the assumption that because each type of twin would be born into and grow up in the same home environment, any difference between the degree of similarity between MZ and DZ twins must be due to genetics because this is the only way that the two sets of twins differ. The same logic would presumably hold for comparisons between other kinship pairs. A variation on this theme was the study of MZ twins reared apart. Because the claim would be that each was raised in a different environment, then any similarities between them must reflect their shared genetics.

Adoption studies have a similar logic. Imagine a child adopted at birth and then raised by the adoptive parents. Then, at some point, measures are taken of the adopted child's developmental status (such as IQ) and then compared with that of the adopted parents. At the same time, any biological children of the adoptive parents are also tested and their scores are compared. Finally, if the original biological parents of the adopted child can be located, they are also tested. The logic would seem to be compelling. If the adopted child's status were more like that of the adoptive parent than that of the biological parent, score one for the environment, but if the adopted child's status is more like that of the biological parent, who has had no contact with the child since birth, then score one for heredity. Actually, it turns out to be much more difficult to assign any score to either the kinship or the adoption studies, but that is for the next chapter's discussion. For now, it is important to recognize that the studies being done in the 1930s were nevertheless quite comfortable doing so.

The intervention and deprivation studies are two sides of the same coin. Deprivation studies withhold (or in the case of humans identify) some experience typical of a species development to see if the development continues in a typical manner even though a significant change has been made in that organism's environment. If it does, then the argument is made that development is not dependent on the environment but rather on some internal mechanism that, short of a total lack of oxygen or food, etc., will proceed pretty much irrespective of circumstance.

Intervention studies identify individuals or a group of individuals whose development is less positive than that of others and attempt to remedy the discrepancy by providing what is seen to be the necessary compensatory experience. If, after some period of time, members of the deprived group are doing just as well as member of the control group, then score one for the environment; if there is little or no improvement in spite of the intentional interventions, then score one, presumably, for genetics.

Kinship and adoption studies became increasingly common by the 1930s because nativists saw them as the perfect vehicle to demonstrate the relative importance of nature over nurture. Intervention studies were favored by empiricists because they provided a way to demonstrate the power of the environment, and deprivation studies, at least with respect to other species, were often used by nativists to show that even in the apparent absence of stimulation, development occurs as it should. Again, the certainty as to one's conclusions evident in these early studies would certainly be seen differently now, but nevertheless we are pretty much doing the very same classic research designs today as was the case almost 100 years ago and, more often than not, reaching the same conclusions on both sides of the aisle. Two examples should suffice, one from each side of the aisle and not, surprisingly, each reaching very different conclusions. First a paper by Leahy (1935) and then one by Neff (1938).

Leahy compared two groups of children. In the first group were children who had been placed into adoptive homes; in the second (control) group, children living with their biological parents. She made the argument, as will be subsequently clearer, that the two home settings were identical across groups. As such, the adoptive children shared only an environment with their adoptive parents whereas the biological children shared both an environment and a heredity with their parents. Therefore, Leahy argued, because the environment in the two groups was identical, any differences between the children in the two groups, measured in terms of intelligence, must be due to heredity.

To accomplish her environmental matching, she identified a control child for each of her adopted children and made sure that the two children matched in terms of sex, age range, father's occupation, father's school attainment, mother's school attainment, residence in a community of at least 1000 residents, and "whose parents were white race, non-Jewish, north-European extraction" (p. 258). This was probably no small task because her total sample size was almost 400 children. In addition, she also determined that the average IQ of the two groups was virtually identical (110.5 for the adopted children, 109.7 for the control children) and that the two groups of parents did not differ in terms of adult measures of intelligence and vocabulary or in terms of a more general measure of environmental status, which included ratings on degree of social participation, available cultural materials, child training facilities, and economic status. At this point, one might conclude that, because the two groups of parents did not differ from each other and the two groups of children did not differ from each other in terms of mean IQ scores, environment was the more dominant influence on children's intellectual development. But as is true of most nativists, the issue for Leahy was not the similarity of the group data but rather indices of individual differences between children, that is, the correlation coefficients between specific children and their specific parents and the patterns of these coefficients that led her to a very different conclusion. She found that on average the correlations between an adopted child's IQ and any one of a number of measures of either the adoptive parent's status or the environmental quality of the home was about 0.20. For the control groups, the average was about 0.50. She concluded that because the two groups of families had been matched on environment, the difference in the two correlation coefficients must be due to heredity:

> Hence variance in intelligence is accounted for by variance in heredity and environment combined to the extent of about 25 percent (square of *r* .50). In the adopted group however, where environment is functioning independently of heredity, variance in intelligence is accounted for by variance in environment only to the extent of about 4 percent (square of *r* .20).... Apparently environment cannot compensate for the lack of blood relationship in creating mental resemblance between parent and child. Heredity persists. (p. 284)

A good summary of the research supporting an empiricist position is Neff's (1938) literature review paper, "Socioeconomic status and intelligence: A critical survey." Neff outlined three goals for his review: (1) an examination of the precise character of the relationship between socioeconomic status (SES) and measured intelligence (IQ), (2) an evaluation of the techniques used to study the relationship, and (3) to interpret the relationship based on all available data. The paper is particularly useful for our purposes because his review of all the available data includes virtually all of the research paradigms used in the classic debate.

Neff started with what was then and continues to be today a well-documented piece of data, namely, that the higher the SES of the family, the higher the child's IQ. In particular, he notes a consensus across multiple studies of approximately 20 IQ points between children of the "lowest urban social group" and children of professionals. Neff is quick to point out that this spread may have less practical significance than it might appear because, given a coefficient of +0.45 between SES and IQ, knowing a parent's SES allows only a 10% improvement in estimating a child's IQ than would be possible "if we simply guessed" (p. 731). But the fact that the spread does exist and that it is routinely reported across studies of various types using a variety of instruments means that there is something here, as far as Neff was concerned, in need of analysis.

Neff first looks at the IQ test itself and notes the problem frequently reported then, as now, of sometimes significant differences in test–retest scores, especially when the testing interval is longer than 1 year. Given this reported pattern, Neff felt there was little support for seeing the IQ score of a child as a constant; rather, he concluded that "we might discover that IQ is relatively constant just so long as the relation between the individual and his total environment remains relatively the same, but that IQ undergoes change with marked shift in the total situation" (p. 737).

What about the effect of SES itself? Neff looked at several studies of children both in this country and in England who lived in very, very limited circumstances. In England, he reports on studies of both "canal boat children" and gypsy children and in the United States, of children living in the mountains of eastern Kentucky, whose experience he described as "an extremely isolated existence at a very low cultural and economic level, perhaps lower than any other rural group in the country" (p. 739). Finally, he looked at studies involving relocation that affected SES, in particular, "Negro" children whose families moved from the South to New York City. Looking at the general pattern across studies, Neff concluded that SES had a significant effect on IQ. Children who remained in very impoverished circumstances had IQ scores that declined over time, reflecting, according to Neff, the cumulative impact of SES. At the same time, significant changes in SES that were due to relocation, as was the case of those moving from the rural South to the industrial North, correlated with a significant positive change in children's IQ scores.

A second way that Neff examined the impact of SES on IQ was through an analysis of children in foster care and adoption. Because foster care and, even more so, adoption were presumed to offer a better circumstance for children than their original homes, the question was whether the improvement in living circumstances correlated with an improvement in IQ score. Here he reports on data that in some ways are at the very heart of the classical debate. He notes that, on the one hand, the correlation (i.e., a measure of rank-order agreement) between foster or adopted children and their biological parent is in fact modestly higher than with the foster or adoptive parent *but*, on the other hand, the actual raw IQ scores are more similar to those of the foster or adoptive parent than those of the biological parent. Then, as now, both sides of the classical debate take comfort in the foster care and adoption studies because, to nativists, the rank-order correlations are seen as proof of the impact of genetics, and, for the empiricists, who note the significant IQ gains resulting from the intervention, the sense is who cares about the rank-order correlations if most, if not all, of the children are doing better than would have been the case otherwise.

In summarizing his findings, Neff was very clear about his conclusion:

> In other words, we have tried to show that although individuals at birth may differ in native endowment (a still incompletely settled question), it has definitely *not* been proved that social status of the parent has anything to do with the native endowment of the infant. That a positive relationship later in life may develop is hardly denied. But all the summarized studies tend to show that low cultural environment tends to *depress* IQ approximately to the degree agreed as characteristic of laborer's children, and that a relatively high environment *raises* IQ correspondingly. All, then, of the twenty-point mean difference in IQ found to exist between children of the lowest and highest status may be accounted for entirely in environmental terms. (pp. 754–5)

Aside from the merits of the two articles and aside from their methodological and statistical "challenges," it is still nevertheless amazing that, on the one hand, Leahy can find data that lead her to conclude that only 4% of the variance in IQ scores reflects home environment whereas, on the other hand, almost simultaneously Neff can conclude that major shifts in IQ scores may be accounted for entirely in environmental terms. It is incredible when you think about it.

Two factors perhaps served to, if not end the classic nature–nurture debate of the first 30 or 40 years of the twentieth century, then at least put it in remission. The first was the political events taking place in Europe in the 1930s that eventually led to the events of World War II. Hitler's racist doctrines and the Holocaust that followed from these doctrines quickly led to the end or suspension of discussions of certainly eugenic thought but more generally of any nativist sentiment (Gillette 2007). At the same time, behaviorism was emerging as the dominant theoretical influence in American psychology. Although it is always hard to pinpoint a reason for significant paradigmatic shifts, the rise of behaviorism, both in the traditions of Skinner (1938, 1953) and of both Hull and Spence (Hull 1943; Spence 1956) in the 1940s, and its dominance through the 1950s and 1960s probably, to one degree or another, reflected Americans' growing sense of ourselves as the society that really can do anything it sets its mind to doing. After all, many no doubt believed that it was the United States that virtually alone won the war and saved democracy. This positive attitude no doubt was also reflected in the growing civil rights movement of the period. All of these events together seemed to lead us as a society to increasingly look at what could be rather than what is, that is, to increasingly look to nurture rather than to nature and to spend less time concerned with sorting and classifying what were seen as stable and enduring traits and capacities and more time creating the social structures that were now seen as creating equal opportunity for all. The Great Society and the great intervention projects such as Project Head Start clearly reflected this optimism, this belief that the inequities in a society do not reflect genetics but rather reflect social structures that existed to benefit the few at the expense of the many.

Within psychology in general and developmental psychology in particular, there was an increasing emphasis on experimental research (McCandless and Spiker 1956) and on understanding the laws of learning. In summarizing what he saw as the "Age of Learning Theories" from 1935 to 1965, Spiker (1989) noted that

> The goals of the major learning theorists were approximately the same, although their theories differed in many important ways. Hull was the most articulate with respect to a theoretical program. He considered classical and instrumental conditioning to be the simplest of all learning situations. He wanted to begin with the laws of conditioning as the axioms (postulates) in the theory of behavior. These axioms included principles describing the building up of associations (habit), the generalization of associative strength, the building up of inhibition, the generalization of inhibition, the development of motivation (drive), and the interactions of associations and motivation. (p. 87)

Perhaps this optimism on the role of the environment in defining development was no better exemplified than in Baer's (1970) paper offering an "age-irrelevant concept of development." Baer argued that studies comparing the performance of children of different ages led to what for him was the erroneous conclusion that these differences reflected the cumulative impact of developmental processes, which in turn tended to correlate with chronological age. But, he argued, if it can be shown through some experimental manipulation that the behavior of a younger child can now be equal to that of an older child, then the experimental manipulation has produced a developmental change in the younger child. For Baer, these experimental manipulations were Skinnerian in form – patterns of reinforcement, punishment, extinction, differentiation, and discrimination:

Consequently, it seems thoroughly reasonable to have built a concept of development squarely upon that technology. The concept that results – to summarize it for the final time – is that to produce certain behavioral change, the procedures of learning technology, which ordinarily work in isolation, in this case may be effective only if applied in a correct sequence to a well chosen set of behaviors. There must be many such sequences for any specified behavioral outcome; and the sequences are not intrinsically lengthy in time. Thus, development is behavior change which requires programming; and programming requires time, but not enough time to call it age. (p. 245)

Although the behavior revolution was certainly redefining the developmental landscape and was increasingly reflected in social policy and practice, not everyone was willing to jump from the extreme of nativism to the extreme of empiricism. There was still some attempt to define the middle ground. The most notable of these was by Anastasi (Anastasi and Foley 1948; Anastasi 1958) whose 1958 paper titled "Heredity, environment, and the question how?" is still considered a classic in the literature and today might well offer a roadmap for the resolution of the new debate. Her concern, however, was problems she saw with what we recognize as the classic debate.

The 1948 paper Anastasi wrote with Foley examined the three different ways that the nature–nurture relationship had been construed. In the first case, which she referred to as isolated operations, heredity and environment were each seen as exerting independent effects on behavior. Such "main effect" theorizing would claim that people inherit specific traits or capacities that express themselves irrespective of environmental circumstance or that a particular environmental effect is so powerful as to exert a uniform influence irrespective of genotype. A second way is to assume that the effects of heredity and environment both contribute to some behavioral outcome in an additive fashion. Such a perspective leads to efforts to apportion variance between heredity and environment. They find that both of these perspectives lack any significant support and that both are inconsistent with the empirical evidence. The only accurate way to portray the relationship between the two, they said in 1948, was to see them as interactive: "According to this view, hereditary and environmental influences, however conceived, are regarded as mutually interacting factors in all behavior, the nature and extent of the influence of each type of factor depending on the contribution of the other" (p. 240).

From this strong interactive perspective, they argue that statistical efforts to disentangle the two are inappropriate because the forms of these interactions are individual specific. Further, claims that differences between individuals would ultimately reflect only genetic factors under "optimal environmental" conditions are equally inappropriate because "the optimum environment for the attainment of any given result will differ for each individual, unless it is assumed that behavior development is independent of any hereditary individual differences. If, however, the definition of optimum environment assumes the absence of such hereditary differences, one obviously cannot conclude that in such an optimal environment hereditary influences are at their maximum!" (p. 243). Their conclusion in this first paper is that we would be better served in understanding development to focus on structure–function relationships rather than in trying to make claims for the relative importance of hereditary and environment. In particular, they argue that because the effect of heredity is expressed through biological structures, the effect of heredity on behavior is always indirect and always in interaction with the environment.

The 1958 paper further elaborates on the ways that heredity and environment might interact. Having made the argument in the 1948 paper that the proportional contributions of heredity and environment to any given trait are not constant but rather vary under different genetic and environmental interactions, she now offers a concept of *how* these interactions might take place. In essence, the more extreme the particular genetic or environmental factor, the less likely that the other will be able to offset the influence. As either becomes less extreme, then the potential impact of the other becomes greater. Further, the form of each interaction then determines the potential forms of subsequent interactions. And all of this exists within a sociohistorical context; for example, the development of new treatments for genetically based diseases means that what was once a situation in which genetics defined virtually all of the developmental variability, now, with the new treatment, the environment comes to play an increasingly significant role in determining variability. It is important to note that her reference to the role of context is significant because typically the classic debate was seen as virtually ahistorical, that is, that the interplay of nature and nurture was a constant unaffected by time and place. Her conclusion is worth noting:

Hereditary influences – as well as environmental factors of an organic nature – vary along a "continuum of indirectness." The more indirect their connection with behavior, the wider will be the range of possible outcomes. One extreme of the continuum of indirectness may be illustrated by brain damage leading to mental deficiency; the other extreme, by physical characteristics associated with social stereotypes. Examples of factors falling at intermediate points include deafness, physical diseases, and motor disorders. Those environmental factors which operate directly upon behavior can be ordered along a continuum of breath or permanence of effect, as exemplified by social class membership, amount of formal schooling, language handicap, and familiarity with specific test items. (p. 206)

So as the decade of the 1960s came to a close, a couple of conclusions would seem to have been in order. First, what started out at the beginning of the twentieth century as a seemingly logical and philosophical debate on the relative, independent roles of nature and nurture became increasingly by the 1930s to be a data-driven debate that very much resembles, in terms of both research designs and findings, the results reported by reductionist nativists and empiricists active today. Second, notwithstanding the view of those like Anastasi, the ascendance of behavior models and social policy favoring a strong environmental message by the late 1960s seemed to be complete and permanent. No one apparently mentioned this to Arthur Jensen, however.

In 1969, Arthur Jensen (1969), published an article titled "How much can we boost IQ and scholastic achievement?" It was not the first time he had addressed this issue because, in fact, he had been a prolific publisher of articles on the interplay of social class, race, intelligence, and genetics for many years. However, this article was different for it set off a firestorm of reaction virtually unprecedented in academic settings. He was fiercely challenged in academic journals, and his public presentations on these topics invariably led to protests and disruptions so serious that it often required the police to restore order. Perhaps only the reception that E. O. Wilson received to his introduction of sociobiology could be considered comparable (Segerstrale 2000).

What was so special about Jensen's 1969 article? Two things probably. The first was the timing. The year 1969 also marked the emergence of a very critical report on the effectiveness of Project Head Start done by the Westinghouse Learning Corporation. The report concluded that a summer-only session had no effect on children's academic achievement and that full-year programs had a limited and apparently temporary benefit because the gains seemed to disappear by the second or third grade. Hardly a glowing recommendation for the cornerstone of the War on Poverty. The second factor that made the 1969 paper such a sensation was where it was published. Rather than again using the respected by admittedly limited visibility of academic journals within traditional psychology, Jensen chose to publish the 1969 paper at the ground zero of the eastern, liberal, politically correct mothership: *The Harvard Educational Review.*

It is important to consider what in fact Jensen did and did not say and also to consider the responses published in the volume at the end of his chapter because the article marks the resurrection of the classic debate, which had been dormant since the beginnings of World War II and as such marks the transition from the historical study of the classic nature–nurture debate to its contemporary study.

The article itself is very well written and probably makes one of the clearest arguments for a nativist, reductionist approach to the classic nature–nurture debate. The article is not the least inflammatory in its tone and provides what Jensen argued is compelling data to support his position. It was more what he came to say as the article developed that created the firestorm. Jensen's aim was to offer an explanation as to why these early compensatory education programs failed to narrow the achievement gap between "minority" and "majority" children. He believed that the primary problem is that the theory on which these compensatory programs were based is incorrect. He describes this "incorrect" theory as having two components: "the average child concept" and the "social deprivation hypothesis." The average child concept, according to Jensen, makes the claim that virtually all children are very much alike in their mental development and therefore are all capable of learning. Further, any variability in learning is therefore a reflection of "rather superficial differences in the children's upbringing at home, their preschool and out-of-school experiences, motivations and interests, and the educational influence of their family background" (p. 4). The second concept, the social deprivation hypothesis, is an extension of the first and, according to Jensen, makes the claim that for those children who are significantly disadvantaged with respect to their peers in school achievement, the best explanation lies with the variables mentioned in the preceding quote, that is, environmental circumstances, and that the most appropriate intervention is to provide early on compensatory educational experiences: "The chief aim of preschool and compensatory programs, therefore, is to make up for these environmental lacks as quickly and intensively as possible by providing the assumedly appropriate experiences, cultural enrichment, and training in basic skills of the kind presumably possessed by the middle-class 'majority' children of the same age" (p. 4). In other words, the primary goal of these intervention programs, again according to Jensen's analysis, is to foster academic achievement by raising children's IQ, and the reason that it is not working is because the primary determinant of IQ is not environmental but rather genetic. And then, over the next several pages, he offers an explanation of the concept of measured intelligence (IQ) and the place of IQ in understanding student achievement. He notes the origin of intelligence testing as an effort to assess student capability, talks about crystallized and fluid dimensions of intelligence, and offers a support for intelligence as a unitary, generalized construct. In making these points, he is also quick to point out that it should be no surprise that there

generally are significant associations between measures of intelligence and school achievement because those measures were designed to predict those who would succeed in a school setting that "evolved within an upper-class segment of the European population and thus were naturally shaped by the capacities, culture, and needs of those children whom the schools were primarily intended to serve" (p. 7). And, by extension, there should be no surprise that there are also significant correlations between measured intelligence and occupation as educational attainment is one of the prime correlates of occupational status. In other words, some occupations have more status because there are fewer individuals, according to Jensen, who have the competency to fulfill them successfully:

Most persons would agree that painting pictures is more satisfying than painting barns and conducting a symphony orchestra is more exciting than directing traffic. We have to face it: the assortment of persons into occupational roles simply is not "fair" in any absolute sense. The best we can ever hope for is that true merit, given equality of opportunity, act as the basis for the natural assorting process. (p. 15)

And, according to Jensen, this natural assorting process is largely genetic in source, a conclusion, he argues, that is confirmed by kinship studies. Further, the significant correlation between SES and IQ, also according to Jensen, has a significant genetic component because mild subnormality in the absence of specific neurological symptoms is virtually confined to the lower social classes: "It is therefore most unlikely that groups differing in SES would not also differ, on the average, in their genetic endowment of intelligence" (p. 75).

Having made the argument of a strong association between SES and IQ, Jensen now turned his attention to what evidence there might be for a association between race and IQ. His argument rested on evidence that distinct populations have different distributions of gene frequencies for different traits, presumably including intelligence. As such, he noted that even when controlling for SES, "Negroes" on average test about 11 points lower than whites. And, he argues further, given the possibility that there is evidence to support the hypothesis that there is an inverse relationship between SES and birth rate, he raises the specter of possible dysgenic trends, in particular that our national IQ might actually be decreasing. And then, in an interesting twist on Galton's eugenic arguments, Jensen asks this question:

Is there a danger that current welfare policies, unaided by eugenic foresight, could lead to the genetic enslavement of a substantial segment of the population? The possible consequences of our failure seriously to study these questions may well be viewed by future generations as our society's greatest injustice to Negro Americans. (p. 95)

Whereas Galton's eugenic arguments were justified in terms of what was presumably best for the ruling class, now Jensen seemed most concerned about what is best for the presumed underclass. And, given his arguments, it is no surprise then that Jensen saw the evidence for little impact of intervention programs as something quite expected because such environmental interventions can have little impact on a characteristic that is primarily defined by what he sees as genetic endowment.

As you might imagine, Jensen's argument did not go unchallenged. In fact, the editors of the *Harvard Educational Review* asked seven scholars to respond to the article and then provided Jensen an opportunity to have a rebuttal.

The critiques of Jensen clustered around four issues. The first is a variation of the is–ought fallacy in philosophy. Hunt (1969), for example, argued that any evidence of group differences must always be situated within context rather than seen as absolute and ahistorical. Hunt's response, which asked if compensatory education had even been tried, made the point that findings of a high heritability for a particular measure say as much about environment as they do about genetics and, in particular, they in no way deny the possibility that if the environment changes, then the heritability of that measure will likely decrease. Hunt asks why anyone would think that a year of Project Head Start, no matter how well done, would be able to offset the cumulative cultural impact of decades of slavery followed by decades of racial segregation and intolerance and/or decades of socioeconomic isolation and impoverishment.

Kagan (1969) made the point that Jensen's focus on measures of individual rankings and their correlates is a much less important indicator of development than average group differences. Even if individuals maintain their relative status, isn't it more important to consider the significant average gains that intervention programs have been found to produce? So what if the rank orderings stay the same if now, as a result of an appropriate intervention, all individuals easily demonstrate competence in some endeavor? Kagan recognized that genetic factors may in fact be most dominant when looking at extraordinary competence in sports or music or mathematics, but such unique situations are almost certainly defined by a different interplay of nature and nurture than the more everyday scholastic and occupational skills that, Kagan argued, virtually all of us are more than able to master.

It is important to take special note of this issue Kagan raises. The issue of rank order versus average differences is a major difference in how nativists and empiricists measure the influence of nature and nurture, respectively. And the fact that the two measures are independent of each other no doubt explains why contradictory findings have been and continue to be reported in the literature. Simply put, the participants in the classical debate have never been able to agree on a dependent variable to measure. For nativists, it is always rank order; for empiricists, group mean differences.

Both Elkind (1969) and Cronbach (1969) argued that the environment gets short shrift in Jensen's analysis. Elkind noted that for Jensen, environmental influence is little more than what is left once each of the *six* measures of genetic variance is removed. Should anyone then be surprised when the environment is found to account for little variance? Cronbach noted that treating the environment as a unitary force is a misrepresentation of the way environment works. Environments cannot be ranked from good to bad but rather must be considered in terms of the degree of fit they offer each individual. As such, there should be no reason to assume that a homogeneous program such as Project Head Start should prove to have equal benefit for all children.

Finally, Hunt, in raising the question of the validity of a reductionist view of either nature or nurture, makes the argument that rather than viewing biological and psychological variance as separate and antagonistic, all the biopsychological evidence points to their mutual interdependence at all levels of analysis. In other words, for Hunt, it was time then to end the classical debate and focus on the new debate.

In 1869, Galton proposed to show that what we are is largely a matter of genetics; 100 years later, Jensen makes the same argument. Over this 100-year interval, both nativists and empiricists, each using their own respective logic, methods, and statistics, claimed to show the strength of their position. And after trying to do so for this 100-year period, neither seems to have convinced anyone on the opposite side of the debate to change their minds. Maybe it is also true that those who *can* remember the past are also still condemned to repeat it.

3

The Proxy Debate

A Primer on Methodology and Analysis

Caveat Emptor – let the buyer beware. Good advice for almost anything from buying cars or houses or making investments and, for that matter, in trying to make sense out of the nature–nurture debates. As is true about anything within the academic disciplines, the nature–nurture debates rest on certain basic assumptions about theory, methodology, and analysis. What is the subject matter? How is it defined and measured? How are data collected? How are they analyzed? How are they interpreted? These are not, in fact, questions unique to the nature–nurture debates but rather are generic to all aspects of the study of human development. Nevertheless, the nature–nurture debates often serve as the primary battleground for skirmishes between opposing sides on each of these issues. It is in this sense that I define one of the nature–nurture debates as the "proxy debate."

This chapter is mostly about the proxy debate because it is these questions and assumptions on which they rest that speak to the heart and soul of human development as a discipline. These questions and assumptions can perhaps best be grouped together in two, admittedly overlapping, broad categories. The first refers to issues related to the question of the most appropriate level of analysis one should use in the study of human development, and it is this issue that serves to differentiate the classic debate from, for the most part, the new debate. The second broad category relates to a host of issues concerning research design and analysis, and it is this issue that most often serves to put the two sides of the classic debate at odds with each other.

The First Issue: The Level of Analysis

The question of the most appropriate level at which to study human development refers to what the philosopher of science Stephen Pepper (1961) defined as a matter of worldviews, in this case, a distinction between a *mechanistic worldview* and an *organismic worldview.* As described by Pepper and others (Reese and Overton 1970; Overton 1984; Goldhaber 2000), mechanistic worldviews are reductionist in perspective and organismic worldviews are systemic in perspective. A reductionist perspective argues that it is possible, theoretically at least, to disentangle the many variables that might influence a particular behavior and examine the relative influence of each variable on the outcome behavior independently of the influences of all other variables. The presumption is that all of the antecedent variables, taken together, constitute 100% of the explanations, or, more commonly, variance, that influence the behavior in question. Relative influence is accomplished by *partitioning* the variance, which in turn reflects both the particulars of the research design and the methods of data analysis. A systems perspective would make no such argument because individual antecedents acquire meaning only when in interaction with each other.

It is possible to distinguish the two worldviews by noting that mechanism is essentially an inductive perspective, working from bits and pieces of data that are seen as coming together, much as a puzzle is put together. Eventually the whole picture is revealed. Organicists, on the other hand, tend to be deductive theorists. They start with the big picture and then look to see how the parts have come to fit together and interact with each other. As such, it should not be surprising that mechanists tend to see themselves, and especially their methodology, as atheoretical whereas organicists see a close interplay between both methods of data collection and data analysis and theory.

Overton (2003) draws a sharp contrast between systems theorists and population behavior geneticists (i.e., the nature side of the classic debate). He sees the behavior geneticists as operating without a clearly articulated metatheory and therefore portraying themselves as working within an "assumption-free scientific method aiming to simplify complexities and arrive at the partitioning of variances into one component labeled 'heredity' and one labeled 'environment'" (p. 356). In contrast, Overton (2003, p. 356) describes the systems theories of the organismic worldview as offering an epigenetic image of development in which systems are transformed: "novel features and novel levels of functioning emerge and these cannot be reduced to (i.e., be completely explained by) earlier features."

A mechanistic worldview rests on five basic metatheoretical assumptions. The first is that behavior and behavior change are natural, lawful, universal, knowable phenomena. And, like any other universals, natural phenomena are, theoretically at least, fully predictable, given enough opportunity for study. The second assumption is that it is possible to use objective research strategies to study behavior and behavior change and that because these strategies are objective, they do not have any more effect on or define the behavior of interest than does the use of a ruler in measuring one's height. The third assumption is that all causes of behavior and behavior change can be characterized as either *efficient* or *material* in nature. Efficient causes are those that are external to the individual; material causes refer to the structure or makeup of the individual. In more practical terms, within the classic debate, material causes are what nativists focus on (e.g., genotype), and efficient causes are what empiricists focus on (e.g., parenting strategies). In the fourth assumption, mechanists believe that the objective study of behavior and behavior change makes it possible to examine the relative influence of each material and/or efficient cause independently of all others. This reductionist belief is the bedrock of the mechanist's approach to the study of behavior and behavior change. Finally, because the study of behavior and behavior change is seen as best approached through a reductionist perspective, in which the influence of each antecedent is added to all others to eventually, theoretically at least, account for 100% of the variance, the process of behavior change over time is best seen as a quantitative process involving the increasing complexity of a set of basic elements common to all age groups (Goldhaber 2000).

It is important to note that a mechanistic worldview has been the dominant approach within psychology almost since the inception of the discipline and has historically reflected psychology's strong desire to move away from introspective research strategies and toward what are seen as objective, quantitative ones, ones most closely resembling, in theory at least, the physical sciences. This view of the process of inquiry as objective is what leads them to often see themselves as atheoretical researchers rather than as being tied to a set of metatheoretical assumptions. In other words, mechanists typically argue that their research simply reflects the dictates of the scientific method, and, as such, their reductionist methodology can be used to test theoretical assumptions reflective of those of any theorists within any worldview. So, presumably, mechanists would see their partitioning methodology as equally appropriate to study the "antecedents" of Piaget's concept of object permanence in young infants as they would the study of the relative influence of nature and nurture on school achievement. This is a very different view than that of the organicists, who clearly see method and theory as interdependent (Overton 2003).

An organismic worldview, on the other hand, eschews reductionist models as inconsistent with what organicists see as the real nature of human development. For these theorists, development can be approached only from the study of the interplay of variables that influence, in a synergistic matter, the course of development over time. Any attempt therefore to partition this synergy would then make it impossible to study development because development, from an organismic perspective, exists only at the synergistic or systems level. Arguing which antecedents, independent of each other, influence a particular behavior, would therefore, to organismic theorists, make no more sense than arguing whether hydrogen or oxygen contributes more to the "wetness" of water.

The organismic worldview posits four metatheoretical assumptions. First, the process of development is seen as best represented as a qualitative rather than a quantitative process, one involving the progressive, active construction and reconstruction of levels of organization, each level or system involving the interplay of components not necessarily found at previous levels. Second, the process of development is seen as a universal, species-specific characteristic of all humans. We may not be all clones of each other, but no one is going to mistake one of us for an aardvark at any point in either of our life spans. Third, the process of development is seen as directional, involving a series of syntheses, each leading to a greater potential for effective adaptations to life experiences. Finally, individuals are seen as having an active role in directing the course of their own development, largely through the process by which meaning is attributed to life events. In contrast to a mechanistic worldview, in which the individual is seen as the more or less passive product of material and efficient forces, here the individual is seen as having the potential to transcend these two influences by a reflective process of giving meaning to life events. One significant implication of this last assumption is that the possibility of full research objectivity and complete knowledge of the causes of development, from an organismic perspective, are questionable because what becomes important to know is not only the particulars of the experience itself – presumably fully knowable – but also the meaning the individual attributes to that experience (Goldhaber 2000).

Worldviews are not, in and of themselves, provable or disprovable. Rather, it is the hypotheses consistent with the premises of each worldview that are testable. Presumably, if enough hypotheses based on a particular worldview are found wanting, then that worldview itself becomes questionable. It would be nice to be able to report that there are studies that directly compare predictions from the two worldviews, but this is not, and probably cannot be, the case. How could it be otherwise when the two perspectives cannot agree on the level of analysis. As such, mechanists tend to criticize organicists over what they would see as a lack of methodological rigor in organismic research, and organicists critique mechanists by arguing that it really does not make any difference how rigorous you are in your method and analyses if your approach denies the very phenomenon you are trying to study. Suffice it to say that neither critique has had much influence on the other.

Perhaps the best way to see how these metatheoretical assumptions are (or are not, as the case may be) reflected in the debates is to start by considering the equation typically used by nativists to identify all of the components that go into defining the phenotypic variance of a population with respect to some defining characteristic of that population. Intelligence tests scores or some measure of personality are often the focus of such research, but in fact any characteristic showing some degree of variability would be equally appropriate. By considering the phenotypic variability, the equation is attempting to identify the relative, additive contributions that several antecedents contribute to the distributions of scores in this population for the characteristic in question, that is, why is it that not everyone gets the same score on the test? The debates about the validity of the assumptions on which the equation is built correspond to the two fundamental issues identified at the start of this chapter, namely, the question of the most appropriate level of analysis and the question of research design and analysis within levels.

Let me first define the terms in the equation and then talk about the issues related to each. Here is the equation:

$$V_{BP} = V_{GA} + V_{GNA} + V_{ES} + V_{ENS} + V_{GEcorr} + V_{GEint} + V_{ERR},$$

where V_{BP} refers to the phenotypic variance in the population on the characteristic of interest (e.g., I.Q. or personality or whatever else is being measured). Theoretically this number should sum to 100% because the seven antecedents on the right-hand side of the equation are seen as accounting for all the variances affecting this characteristic.

V_{GA} is defined as *additive genetic variance* and is seen as reflecting the direct, linear contribution of the genes we inherit from our biological parents, or, as Jensen put it, additive genetic variance responds to selection according to the simple rule that "like begets like" (Jensen 1969). Additive genetic variance is seen as most closely resembling Mendelian patterns of inheritance in which it is assumed that alleles for a particular gene vary in a quantitative fashion. The science behind animal breeding reflects the importance of additive genetic variance. Presumably, the best way to produce a fast race horse is to mate a fast stallion with a fast mare.

V_{GNA} refers to *nonadditive genetic variance* and reflects the fact that not all genes act independently, but in some cases the expression of these genes reflects interaction with other genes at other loci in a nonlinear fashion. Being able to predict the expression of genes that act in a nonadditive or *epistatic* fashion is more difficult simply because they do act in a nonlinear fashion. One doesn't always get a faster race horse, and even when one does, the horse may come with some less than desirable side effects.

V_{ES} is defined as *shared environmental variance* and is seen as reflecting the degree to which individuals sharing the same physical and social environment are affected equally by that environment. Research looking at the effect of parenting styles on children's development, for example, is looking at the effect of shared environmental variance. Shared environmental variance is seen as making people who share a common environment more like each other. Shared environmental variance is sometimes also described as *between-family variance* because it as seen as creating differences between families

In contrast, *nonshared environmental variance* (V_{ENS}), also referred to as *within-family variance,* is seen as the means by which people come to be less like each other even though they do share the same physical and social environment. Differences between siblings raised in the same household are often seen as reflecting nonshared environmental variance, especially when these differences become greater over time.

The next two terms in the equation, V_{GEcorr} and V_{GEint}, each refer to the potential interplay between genetic and environmental variables. In the first case, the *gene–environment* (GE) *correlation,* genes and environment are seen as covarying as might be the case when, for example, the child of two gifted musicians is seen as acquiring from the parents presumably not only genes fostering musical talent but also, most likely, a home filled with musical instruments, frequent musical activity, and more generally a love of all things musical. In such a case, it would be no surprise that the child also shows some degree of musical talent, but determining in what sense and to what degree this talent is genetic or environmental is problematic. One reason nativists like adoption studies so much is that they are considered as potentially eliminating this confound.

In the second case of V_{GEint}, *gene–environment interaction,* the presumption is that there is an interaction between genes and environment such that different genotypes react differently in different environments and vice versa. As such, it is impossible, if such interactions are present, to show a direct relationship between either genotype and behavior or environment and behavior because the expression of the genotype is contingent on the particular values of the environment and the impact of the environment is contingent on the particular characteristics of the genotype it is interacting with. Each is dependent on the status of the other. One way to appreciate the distinction between these two terms is that when genes and environments are seen as covarying, the two components are nevertheless seen as combining their influences in essentially an additive manner; in the case of true interactions, there is no assumption of additivity because the value of each is contingent on the value of the other. In fact, it is the issue of the interaction term in this equation that it the primary battleground between reductionists and systems theorists.

It is one thing to argue that theoretically all of the variance can be accounted for by these six variables; it is another thing to document it in practice. There is always some degree of measurement error present in any research, there is always the possibility that the research instrument itself does not fully capture the impact of what is being measured, and there is always the possibility that some unidentified factor is also contributing to the variability in the expression of the behavior. Therefore the equation also includes an "error term" (V_{ERR}). The amount of variance attributable to error is determined by subtracting all of the variances attributable to the first six variables from 100%. The remaining percentage is the percentage attributable to error. Clearly, the larger the error term, the less confidence one has in the findings.

The formula seems straightforward enough: What could possibly be points of controversy? Actually, every point is. With respect to the first question, the question of levels of analysis, the controversy focuses on if, as systems theorists argue, all developmental antecedents function in an interactive fashion, then there is no legitimate basis to look for additive main effects. With respect to the second question, the one dividing the two sides in the classic debate, the controversy focuses on issues of within and between family variance, on GE covariance, and even on the very research designs used to actually collect data.

If, as organicists argue, all antecedents influence behavior in an interactive, synergistic, systemic manner, then theoretically there cannot be main effects. But if there cannot be main effects such as additive and nonadditive genetic variance or shared and nonshared environmental variance, then what is being measured in this equation? The answer, for organicists, is that what is being measured are artifacts created by the particulars of the research designs and the mathematical treatment of the data. In other words, the numbers have nothing to do with reality. *Au contraire*, mechanists are quick to respond. These numbers do not reflect some sort of statistical slight of hand but rather simply what is there to be measured.

This sharp difference of opinion is very evident in the comments of those who have been involved in the debate. Chiszar and Gollin (1990), for example, make the point that whereas partitioning models may have once made sense, they are now an "utter anachronisms" and in fact make it actually more rather than less difficult to come to a full understanding of developmental processes. It is worth noting that the "now"' they refer to was 1990. On the other hand, Detterman (1990) argues that statistical methods and theory are independent, that statistics are a "mere tool" that have no more to do with the correctness of a theory of behavior genetics than "arguments over the refining powers of optical and atomic microscopes to theories of cell development" (p. 131).

Gottlieb (2003), taking a very strong systems perspective, argues that development is always the consequence of organism–environment interactions in which the *quantitative* (italics in original) contribution of either cannot be specified. And the reason, for him, is very clear: "this is not a statistical limitation but, rather, logically indefensible" (p. 338). In the same vein, Meaney (2010) argues that the "notion of a main effect of either 'gene' or 'environment' is biologically fallacious. Instead, the development of the individual is best considered as the emergent property of a constant interplay between the genome and the environment" (p. 45).

Contrast both Gottlieb and Meaney's statements with that of Plomin and Asbury (2005), who take a strong position in favor of partitioning main effects, even claiming that in correlational studies of DNA variants and behavior, because the behavior of the individual does not change the genotype, the correlations *can* be interpreted causally, that is, DNA differences cause behavioral variability.

Or consider Hunt (1997), who in making a strong argument for what he sees as objectivity in science makes the point that humanistic arguments[1] "appeal to us because it makes us aware of broad classes of influences on our life, without spelling out the specifics of application" and that comparing such humanistic models with the work of those such as "Newton, Einstein, Feyenmann, or the behavior genetics" (pp. 537–8) is not a contest between scientific models; it is a clash between scientific models and worldviews.

These fundamentally divergent viewpoints can be seen playing out in two related nature–nurture issues, one concerned with the degree to which either genes or environments can limit the expression of the other and the other with whether the presence of gene–environment interactions does in fact preclude the validity of main-effects analysis.

The reductionist argument that both genetic and environmental variables are independent of each other (i.e., that both are main effects in the analysis of variance (ANOVA)) allows for the possibility of each setting a limit on the effect of the other. In practical terms, this notion of limits is more likely to be found in the nativist literature. Genotype is seen as limiting the potential effect of environment. System theorists, denying the independent, additive influence of either genotype or environment, make no claim of limits and

[1] Hunt's use of the term humanistic in this context probably should be equated with what others would refer to as an organismic worldview, especially because Hunt cites the work of Piaget and Freud, two organismic theorists, as humanistic theorists.

instead talk about the potential product of the interaction of the two variables. This distinction is most easily seen in contrasting reductionist's concept of the *reaction range* with systems theorist's concept of the *norm of reaction* (Griffiths and Tabery 2008).

The reaction range (Gottesman 1963) is primarily a statistical way to consider the interplay between nature and nurture. The presumption is that each genotype has a range of expression that in turn reflects the particular environment that genotype finds itself in. These ranges can differ both in terms of absolute level, reflecting the influence of additive genetic variance, and in terms of variability, as would be the case when some genotypes are seen as more open to environmental influence than other genotypes. For example, accepting the argument for additive genetic influences on intelligence, one would argue that the final intellectual competence of the individual would reflect the interplay of genotype and environment but that the range would be limited by genotype. In other words, someone considered to have limited genetic intellectual potential would certainly benefit from an intellectually responsive environment, but no amount of enrichment, the argument goes, is going to make this person a genius. This, in essence, is the heart of Jensen's 1969 argument (Jensen 1969). There is certainly a type of interaction reflected in the reaction range concept in the sense that the individual's intellectual competent is determined by the unique combination of genotype and environment, but because the two are still seen as acting independently of each other and because one, genotype, is seen as setting a limit on the effects of environmental influence, this form of interaction is probably best appreciated as a form of covariance.

A norm of reaction (Gottlieb 2003), on the other hand, does not see either genes or environment as independently limiting behavioral expressions but rather sees each unique combination as producing a unique behavioral outcome, one not fully predictable from knowing other combinations. In other words, unlike the image of intellectual competence within the reaction range, in which the more intellectually enriched the environment, the more competent the individual up to the genotypically defined limit, the reaction norm allows for the possibility that more of a good thing, such as more intense intellectual enrichment efforts, might actually prove to be a bad thing in that at some point the intellectual gains begin to reverse themselves or that even making statements about one's potential are theoretically meaningless because, even if there is now no known environment intervention to offset some genetic effect, this certainly does not mean that there might not be one in the future. In fact, the entire history of medical research is a series of discoveries that have done just this.

But remember that there is also an interaction term that is distinct from the covariance term in our equation. If the reaction range is best understood as a measure of covariance, what does the interaction term in the equation reflect? As with just about everything else in the nature–nurture debates, the answer depends on whom you ask because interactions can be defined in either statistical terms or in biopsychological, systemic terms. Reductionists favor statistical definitions; systems theorists favor biopsychological definitions. As is true of virtually all things when levels of analysis are discussed, the two definitions are not directly comparable nor are they necessarily contradictory. Rather, they reflect the basic differences that exist between those working at an inductive, additive, reductionist level of analysis and those working at a deductive, interactionist, systemic level of analysis.

Guo (2000) describes statistical interactions as referring to the effect of product terms or the heterogeneity of main effects and are thus seen as departures from additivity. The presence or absence of statistical interactions is largely dependent on the scale one uses to measure the effect and the models used to describe or fit the data. In contrast, Guo describes biological interactions as context dependent and existing at one or more levels ranging from the molecular through the cellular, organismic, or population levels. In this case, the presence or absence of biological interactions is largely determined by environmental dependency on gene expression or as a genetic response to environmental change or stress.

This distinction between statistical and psychobiological levels of analysis is also reflected in how genes and gene action are defined (Griffiths and Tabery 2008). At a reductionist level, genes are typically defined in classic, Mendelian terms, that is, as intervening variables in the genetic analysis of phenotypes. In contrast, Griffiths and Tabery (2008, p. 345) note that "many critics of developmental behavioral genetics, especially those with a background in developmental psychobiology, conceptualize genes as determinants of the value of a developmental parameter in the context of a larger developmental system."

For the inductive reductionist, no assumptions need to be made; the data speak for themselves. For the deductive systems theorist, metatheoretical assumptions guide and define how data are gathered, analyzed, and interpreted. The distinction between the two approaches is vividly evident when those reductionists reporting the lack of statistical interactions debate with those systems theorists who find such interactions commonplace in the animal literature. For systems theorists who start with the given that interactions at all levels are the norm in defining the course of development, the reported lack of such interactions in the human data is seen as a reflection of methodological and statistical inadequacies. As Meaney (2010) argues,

> For whatever the difficulties in defining Gene x Environment interactions, research in biology reveals that the genome cannot possibly operate independently of its environmental context. The biological perspective reveals the futility of the nature-nurture debate and of additive models of genetic and environmental influences in defining phenotype. (p. 42)

However, in response to such deductive arguments and in particular in a response to an article by Wahlsten (1990) critiquing reductionist approaches, Plomin, a leading behavior genetic researcher argues that he has tried hard but in vain to find gene–environment (GE) interactions (Plomin 1990b):

> There is no conspiracy against interactionism: If an interactive model can to shown to fit the data better than the traditional model, researchers would be quick to use it. In summary, it is a lot easier to talk about GxE than it is to find it. Rather than trying to shoot the messenger because of his message, it would be far more useful to collect empirical data that demonstrate the importance of GxE. (p. 144)

So, posturing aside, what are the explanations for the discrepancy between the animal and human research? We know Plomin's argument, the interactions simply are not there. The other side of the statistical argument, however, concerns the power of ANOVA techniques to detect such interactions.

All forms of statistical analysis ask the same fundamental question: What is the likelihood that the degree of difference in the data found between two or more groups is likely due to chance alone? If the mean differences between the groups are large enough to be considered statistically significant, then the differences are not seen as being due to chance but rather to the antecedent conditions of the study. But here is the problem. All statistics work on samples drawn from a population rather than from the population itself. The probability estimates are estimates of the size of the effect within the entire population based on the evidence found in the sample. In fact, if it were possible to test the entire population, then we would have no need of any probability estimates because we would, in fact, know if the treatment effects did or did not make a difference. But we do not test populations; we test samples drawn from populations and therefore need to do the statistical analysis to get the probability estimate. The issue (or problem, depending on your point of view) is that the various ANOVA techniques used in research are based on the additive model first developed by Fisher (1918) and as such are not as sensitive to the presence of interaction as they are to main effects. In fact, it would take a sample size many times larger to detect interactions than it would to detect main effects (Wahlsten 1990, 2000). The basis for this sample-size argument is that because interactions tend to be a characteristic of individuals rather than uniform across a sample, the effect of these interactions can be masked in group means. It is not masked in animal data because selective breeding allows for the creation of genetically uniform strains that can then be exposed to differing environments to see whether the behavior outcomes of these pairings is best appreciated as additive or interactive. The majority of the animal evidence shows it to be interactive (Guo 2000; Lewontin 2006; Wiebe et al. 2009; Meaney 2010) and, the systems argument goes, if interactions are present, statistical tests for main effects are not appropriate.

McClelland and Judd (1993) also note two other statistical reasons why interactions are so much more common in animal than human research. First, because human research cannot have the same level of stringent experimental controls than animal research, there is greater error variance in human research, especially when there cannot be randomization of assignment to treatment groups as is the case in all nature–nurture-related research. Greater error variance means that measures of individual variables such as additive genetic variance or nonshared environmental variance are less reliable and this error variance becomes disproportionally greater when the focus is on the measurement of GE interactions. They also note that restricted sample variability also serves to reduce the power of the analysis to detect interactions. This would certainly be the case in adoption designs because adoption agencies are very selective in deciding who is qualified to adopt a child.

As is true for virtually all of the differences discussed in this chapter, we end up with a stalemate between those who favor additive reductionist models and those who favor interactive systems models. In terms of nature and nurture, it is the distinction between the classic debate and the new debate, but more generally it is a distinction about the interplay between how one conceptualizes the study of development and how one collects and interprets data for the purpose of drawing conclusion about development. What we end up with in trying to resolve the levels of analysis question is what Turkheimer (2000) refers to as a "gloomy prospect":

> If the underlying structure of human development is highly complex, the relatively simple statistical procedures employed by developmental psychologists, geneticists, and environmentalists alike are badly misapplied. But misapplied statistical procedures still produce what appear to be results. Small relations would still be found between predictors and outcomes, but the underlying complex processes

would cause the apparent results to be small, and to change unpredictably from one experiment to another. So individual investigators would obtain "results," which would then fail to replicate and accumulate into a coherent theory because the simple statistical model did not fit the complex developmental process to which it was applied. Much social science conducted in the shadow of the gloomy prospect has exactly this flavor. (pp. 163–4).

Caveat Emptor.

The Second Issue: The Focus of Study

The first issue was all about levels of analysis. There is no obvious resolution to this dispute because there is no research strategy that everyone would agree on to test competing hypotheses consistent with the organismic and mechanistic worldviews. Therefore we have a situation in which each pursues research within his or her preferred level of analysis, arguing that the information gained through such efforts is valuable in helping us better understand the human condition.

But when it comes to the second issue, everyone in the classic debate at least, that is, both nativists and empiricists, does agree on both the level of analysis – reductionism – and even on the basic methodology. The disagreements here are (a) over a focus on stability or change, (b) what constitutes a genetic or environmental variable, and (c) the primacy of genetic and environmental measures. Let's take each in turn.

Stability and Change: When Cronbach (1957) and then McCall (1981) each rallied against the two worlds of psychology, they were talking about the gap between those who, on the one hand, see the study of human development as the study of those stable, enduring qualities that distinguish individuals and, on the other, the study of the common mechanisms of change through which individuals grow and adapt. In terms of the proxy debate, nativists and, to some degree, even evolutionary psychologists strongly favor a stability perspective and empiricists and, again, to some degree, systems theorists favor a change perspective. The fact that a focus on stability tends to align itself with nature and a focus on change with nurture is not a mere coincidence but it is also not a necessity. Nevertheless, in practical terms, this alignment is the typical pattern. For behavior geneticists, the focus is on documenting the role of genetics in determining individual differences. The presumption is that for the most part these individual differences are seen as relatively permanent characteristics of individuals; otherwise, why spend so much time and effort documenting them or talking about their social and political significance? Evolutionary psychologists are also interested in enduring, stable, inherent qualities, but their focus is at the species level; for them, the individual differences that behavior geneticists study are trivial compared with the impact of our shared evolution as a species over thousands and thousands of years (Cosmides and Tooby n/d). For the evolutionary psychologist, the focus is on how evolution has shaped the particular characteristics that are shared by all members of the species, characteristics that are often evident even in newborns (Spelke and Kinzler 2009). This difference in focus between the behavior geneticists and the evolutionary psychologists helps to explain why the focus of the former tends to be on issues related to race and socioeconomic status (Goldsmith 1993; Rowe 2005) and the latter on issues related to gender (Pinker 2004).

For empiricists and systems theorists, the focus is on documenting the plasticity or openness of the individual or species to change. Documentation of this openness is done by showing how the introduction of some antecedent condition such as a particular educational approach can significantly influence one or more behaviors. It is not that these researchers argue that human potential is unlimited, but rather than we simply have no sense of what these limits are and, as such, the concept of limits is scientifically not useful. In essence, it is the distinction between the behavior geneticist's concept of the reaction range and the system theorist's concept of the norm of reaction.

Not surprisingly, such contrary views of the human condition lead to different emphases in the study and measurement of behavior and behavior change. In particular, nativists are much more likely to use correlational techniques because these are particularly effective in documenting the ordering of individuals (i.e., measures of individual differences) and empiricists are more likely to use analysis techniques that allow for the demonstration of hopefully statistically significant differences between the means scores of groups receiving different treatment conditions. One interesting thing about the two measures is that they are independent of each other, and, as such, knowing about one tells you nothing about the other. What further complicates the issue is that behavior geneticists typically report only correlation coefficients and empiricists report only mean values. But as Weizmann (1971) noted several years ago, you really need both measures to get the complete picture:

> As McCall's data illustrate, however, there is nothing inconsistent in maintaining that even abilities reflecting a high genetic "loading" [quotes in original] may be quite amenable to change. This interpretation is also consistent with the data obtained from several older studies such as that of Honzig. She reported that, despite high correlations between the IQs of a sample of foster children and their

biological parents, the IQs of the children shifted on the order of 20 points in the direction of the foster parents. (p. 389)

Goldsmith (1993), however, maintains that the issue is not one of reconciling the two measures but rather recognizing that each has value in its own right:

Individual differences are the stuff of behavior genetics, and classic behavior genetic influences are confined to genetic and environmental effects on phenotypic variance, not genes or environment per se. There is no contradiction in analyzing individual differences by linear regression of outcome on sources of variation, even when the individual differences result from highly contingent developmental processes operating in the lives of individuals. In fact, psychologists frequently do analogous exercises. For example, both early IQ and quality of schooling might predict later academic achievement of children in a linear fashion. Computing the relevant regression and interpreting the partial regression coefficients is a legitimate and potentially useful exercise even though the actual learning experiences of the children was highly contingent, interpersonal, and context bound. Analyzing the nature of the contingencies and contextual influences is simply a different task. (p. 329)

Analyzing individual differences to determine the sources of variation is, at a reductionist level of analysis, largely a matter of determining the percentage of variance that is due to genetic factors and the percentage of variance that is due to environmental factors. In terms of genetic influence, this percentage is expressed as a coefficient of hereditability or H^2. One could, in fact, as easily compute an E^2 to express the environmental influence, but this statistic is not computed by empiricists, perhaps because behavior geneticists at least recognize both genetic and environmental influences on development, whereas empiricists, truth be told, rarely even mention the "G" word.

H^2 is at the same time a very simple and a very complex statistic, largely because its significance is so poorly understood and often so incorrectly touted. Because H^2 is expressed as a percentage, it is always a relative term that will vary as a function of the presumed relative influence of both genetic and environmental variances. If the environmental variance for a behavior goes up, H^2 goes down and vice versa. As such, whatever the actual value of H^2 determined in a particular study, that value is specific to that sample drawn from that population; it cannot necessarily be presumed to be an accurate representation of a different sample drawn from the same population or, even more so, of the distribution of variance drawn from an entirely different population, a particularly important consideration where arguments about racial differences are concerned (Sternberg et al. 2005; Whitfield and McClearn 2005). Even as Jensen (1997, p. 43) notes, "Estimates of H are specific to the population sampled, the point in time, how the measurements were made, and the particular test used to obtain the measurements."

This issue of sample specificity helps explain why H^2 might be considered much more useful in animal breeding than in making sense out of human development. In animal breeding, because there is the potential to exert very stringent controls on the environment and therefore reduce environmental variability, the likelihood of getting that fast race horse does go up. Fortunately, the same constraints don't hold for us (Jacquard 1983).

H^2 is a population statistic and cannot be applied to an individual simply because any measurement of an individual, such as one person's height, has no variability; it is what it is. H^2 should also not be confused with cause. To say that the hereditability of some behavior characteristic found in a sample drawn from a particular population is, for example, 0.6 does not mean that 60% of the cause of that characteristic is genetic and therefore 40% is environmental but simply rather that a greater degree of the phenotypic variability of that characteristic, within that sample, is genetic rather than environmental. Ironically, those qualities that we would generally assign significant genetic influence to, such as five fingers on each hand and five toes on each foot, actually have a very low hereditability coefficient simply because virtually everyone has five of each in the right place. In fact, in the case of digits, E^2, if it was to be calculated would actually be a much larger value because almost all such physical malformations are seen as reflecting some problem within the prenatal or postnatal environment (Blumberg 2009). Said another way, evidence of a high hereditability may still say nothing about the modifiability or phenotypic plasticity of that characteristic, a point that Jensen's critics and others have often made (Hunt 1969; Vreeke 2000; Lewontin 2006). And evidence of low hereditability says nothing about the degree of genetic influence on a particular characteristic.

One good example of the contextualized nature of H^2 comes from research looking at the relationship of hereditability estimates for achievement and cognitive competence and socioeconomic status (Scarr and Ricciuti 1991; Scarr 1992, 1993; McCartney and Berry 2009). Behavior geneticists typically report that H^2 is lower in children from low-income communities than children from more economically advantages settings. The explanation given is that the relatively limited opportunities for intellectually enriching experiences in low-income communities serve to dampen children's expressions of their intellectual potential and as such lower H^2. On the other hand, children from more advantaged environments, seen as having all the necessary intellectual opportunities to develop their intellectual potential more fully, show less variability with respect to the influence of environment on cognitive competence and, because there is less environment variance, hereditability increases. It is in this sense that Scarr made reference to "the good enough parent" in her Society for Research in Child Development (SRCD) Presidential Address and that Herrnstein (1973) claimed that as equal opportunity became greater for all in a society, the only explanation of individual achievement would ultimately become genetic, that is, full equal opportunity would reduce environmental variability to zero, making genetic variability 100%.

So what's the big deal with H^2? Why is it such a controversial statistic? Two reasons. Those who advocate for it do not always remember its limitations and those who oppose it do not always remember its limitations. Plomin and Asbury (2005), for example, argue that genetic research has consistently shown hereditable influences in areas such as mental illness, personality, cognitive disabilities and abilities, drug use and abuse, self-esteem, interests, attitudes, and school achievement. Nowhere is there reference to the limits of the generalizability of H^2 that even Jensen acknowledges. At the same time, empiricists, working within a reductionist model, tend to incorrectly equate H^2 with cause and therefore reject it out of hand. To Meaney (2001), a systems level theorist, both miss the mark:

> We have ample reason to celebrate the technical advances associated with the Human Genome Project. Yet, the same technology bears the risk of expanding the divide that lies between the biological and social sciences in the same way that access to computer technology expands the division between the developed and underdeveloped world: one group blindly infatuated with the explanations that might flow from gene technology, the other group huddled in terror at the thought of a biological world of which they know nothing. (p. 51)

Caveat Emptor.

Independent and Dependent Variables: It might seem reasonable that if people are debating the relative merits of nature and nurture that everyone would have a pretty good idea of what each reflects, but this isn't necessarily the case. Behavior geneticists do not make claims as to what specific genes are explaining their correlation patterns; they only argue, for example, that if the correlations between identical twins is higher than those between fraternal twins (and making the assumption of an equal environment for both), then the difference must be genetic. But, in fact, with the exception of those fortunately rare genetic conditions in which a single gene can cause traumatic or even terminal outcomes, for the most part, we simply do not know what genes, in what combinations, are contributing to the variance, that is, we do not actually know what the independent variable is. Again, for the behavior geneticists, this is not that much of a problem because their goal, as Goldsmith

(1994) points out, is to account for phenotypic variance, not to necessarily identify the specific genes explaining that variance.

In some ways, the problem is even greater for the empiricists because trying to get some sort of conceptual handle on as vast an array as everything from the moment of conception onward is no easy task. And in spite of the fact that the vast majority of research on human development focuses on the role of the environment, the truth of the matter is that we really do not have a shared definition of the environment:

> There is no agreement about how to best parse or define the environment in which a person is growing up. There are no standard units of environmental variables, and the definitions of developmentally facilitative environments find no consensus. At best, the efforts to manipulate environmental variables focuses on gross dimensions, at worst, the efforts to define and manipulate the environment are laden with value judgments about good (positive/effective) and bad (negative/ineffective) that have minimal empirical support. (Horowitz 1993, pp. 347–8)

And, as Horowitz argues, if there is no standard definition of or measurement of the environment, then researchers are left to their own devices.[2] For example, one methodological explanation of why behavior geneticists and empiricists see such a different role for the environment is that each measures it in different ways. Behavior geneticists, more likely using larger sample sizes, typically favor self-report or inference based on demographic data. Empiricists, typically using smaller sample sizes, favor direct observation or measurement. So, for example, behavior geneticist's report high IQ correlations between identical twins separated at birth and conclude from such data that because each twin grew up in a different home, the only thing they had in common was their genotype, that is, IQ is largely determined by genes. However, empiricists argue, given adoption policies in the United States and most other Western countries, it is very likely that the two adoptive homes the twins were placed in were actually very similar in terms of socioeconomic status, values, beliefs, location, and religion. As such, even though the two twins did not reside in the same household, they may well have resided in essentially the same environment, and, as such, their high IQ correlations could as much be a reflection of environment as genotype (Wachs 1993; Bronfenbrenner and Ceci 1994).

A lack of standardization not only means that there is no consensus about how or what to measure but for that matter even when to measure, that is,

[2] It should be noted that there have been efforts to systematize the environment, most notably the work of Wachs (2000, 2003) and of Bronfenbrenner (1999, Bronfenbrenner and Evans, 2001). The particulars of each effort will be discussed in the next chapter.

do some environmental effects not become evident until after some passage of time? How do environmental effects, both simultaneous and sequential, influence each other? What should be the appropriate dependent measure of an environmental effect, as parenting strategies, for example, can presumably affect many aspects of a child's and later adult's behavior and do so in different ways at different levels of development?

And to make matters even worse, even if we are able to come up with a clear definition of a genetic effect and an environmental effect, we are still stuck with the problem of demonstrating the independence of the two. The issue involves the first four terms of our equation. The first two, additive and nonadditive genetic effects, are seen as reflecting genetics, and the second two, shared and nonshared environmental variances, are seen as reflecting the effects of the environment. How do nativists and empiricists see the role of each of these four main effects?

In the 1988 movie *Beaches,* Bette Midler's not surprisingly over-the-top character is having a conversation with Barbara Hersey's character. Midler's character goes on and on about herself and at one point stops and says "But enough about me, let's talk about you . . . what do you think of me?" My guess is that empiricists must often feel like Hersey's character when it comes to trying to lay claim to the importance of the environment given the fact that the behavior geneticists argue that not only is there little evidence of gene–environment interactions but that there is virtually no evidence for shared environmental variance, an environmental main effect, and what evidence there is for nonshared environmental variance, the other environmental main effect, is in actuality, according to the behavior geneticists, really a reflection of genes rather than environment:

> The conceptual leap here is to treat environmental measures as dependent measures in quantitative genetic analyses. It is a leap because it seems bizarre at first to ask about genetic influences on environmental measures when environments do not have DNA. However, measures of psychological environments, such as parenting, peers, and life events, are not measures of the environment itself. To some extent, all such measures depend on behavior and can thus reflect genetic influence on behavior. For example, the number of books in the home, a classic item on measures on home environment, is not a measure of the environment per se. Books do not appear on the shelves by themselves; they are put on the shelves by parents. Parents differ on the number of books they put on the shelves and these differences in parental behavior can be caused by parental characteristics, such as liking to read, that may be influenced by genetic factors. (Plomin 2009, p. 64)

How can this be? How can the environment have no independent effect but simply be a function of genetic influence? How can the environment be a dependent variable? The answer requires one to remember that behavior geneticists are interested in individual differences rather than in mean differences. In other words, it isn't that they are saying that parents, and so forth, have no influence on their own children's development; rather they are saying that they have no differential influence and to understand this claim requires a refresher course on research design.

Ideally, a research design within a reductionist paradigm allows for the independent testing of main effects as well as for any possible interaction terms, and doing this requires, again ideally, the random assignment of individual to each of the treatment conditions. Once this is done, and the data from each individual in each group collected, then the data can be analyzed for both main and interaction terms. For example, consider a simple experimental design having two independent variables, teaching method and number of students per classroom, and each variable having two levels (perhaps rote learning versus experiential learning and fifteen versus twenty-five students per class). The 2 × 2 design then allows for the testing of the influence of teaching method independently of class size by averaging across the two class sizes for each teaching method and then doing the same with respect to class size. Testing for the possible interaction of the two variables would involve a comparison of each of the four unique combinations of teaching method and class size.

Sometimes, however, situations are not ideal, especially when it comes to gene–environment research designs. The problem of course is that, unlike in animal research, in human research you cannot randomly assign someone a set of genes and you cannot randomly assign someone a family. The behavior geneticist's emphasis on kinship and adoption research designs is then the best alternative strategy. But at least with the kinship designs, environment and genetics are confounded because individuals share both. The adoption design would seem to solve the problem, especially if information is available not only for the adoptive parents but for the birth parents as well. In such cases, the behavior geneticists report that the correlation between adopted child and birth parent is typically higher than that between adopted child and adopted parent (Plomin and DeFries 1983; Plomin et al. 1993). In fact, they report that the adopted child–adopted parent correlations are typically very low and actually decrease as the adopted child grows older, and, potentially even more damning, is the frequently reported finding that there is virtually no correlation between adopted children and the biological children of the adopted parents, even though all are living in the same family setting (Scarr and Weinberg 1983).

How could this be so if environment makes a difference? The answer, to the behavior geneticists, is that the shared environment does not make a difference, at least as they define difference, that is, explaining phenotypic variability. However, also keep in mind that since correlation coefficients and mean differences are independent of each other, it is typically the case that even if the *rank order* of a group of children's IQs might correlate more highly with those of their birth parents than those of their adoptive parents, at the same time it is much more likely that the actual IQ *levels* of the children are closer to those of the adoptive parents than those of the birth parents. Admittedly the two findings seem contradictory but in fact they are not. The real question is the relative importance of each finding. For nativists, the correlations are more significant because they highlight the importance of genetics; for empiricists, the mean values are more important because they highlight the importance of environment.

So, accepting for the moment the argument that there is virtually no evidence for the shared environment to have a shared influence on those sharing that environment, why might this be so and, by the same token, what then can account for there being much stronger evidence for high levels of nonshared environmental variance (i.e., people sharing the same environment not necessarily being like each other)?

Nurture via Nature: How can it be possible for the behavior geneticists to argue that environmental variables such as socioeconomic status or parenting behavior are best appreciated as dependent rather than independent variables when there are countless studies of both treating each as independent variables? The answer reflects the behavior geneticist's argument that the stronger evidence for nonshared environmental variance is actually an argument for the role of genotype in shaping one's environment. In particular, they argue that an individual's unique behavioral characteristics, which they would attribute to genetics, serve to evoke responses in others and, even more so, as children mature into adults and seek their own lives, that these same genetic factors serve to actively direct the person to find his or her particular niche in life, a niche largely consistent with this person's genotypic characteristics (Scarr and McCartney 1983; Jensen 1997; Rowe and Rodgers 1997; Turkheimer and Waldron 2000; Turkheimer et al. 2009). So, the argument goes, we treat sons differently from daughters not because of culturally defined sex-role expectations but because boys and girls act differently because of their sex-linked genetic differences.

The bottom line for all of the debates between nativists and empiricists working within a reductionist tradition is that no matter how you partition, analyze, control, and interpret the data, you still end up with a chicken-and-egg question. Both nativists and empiricists are trying to identify the unique roles of individual antecedents, and whenever either side makes a claim for one antecedent or another, the other side quickly raises methodological concerns that are seen as compromising or even negating the other's claim. And although the literature is typically more critical of the claims of the behavior geneticists, the fact of the matter is that because both nativists and empiricists share the same fundamental metatheoretical assumptions of a reductionist model, both are equally vulnerable to methodological critiques of claims for the identifications of specific causes of behavior and development. Note that those working at a systems level are less subject to methodological issues because no effort would be made to identify individual causes. The focus instead would be on the understanding of process rather than on the identification of ultimate causes, a much better state of affairs to systems theorists; a not very satisfying state of affairs to reductionists. Ultimately the decision as to the relative merits of the different approaches will come down to the degree to which each helps us to not only understand the human condition but to do what is possible to improve it.

Caveat Emptor.

4

The Classic Debate

This chapter and the next examine the four relatively distinct approaches to the nature–nurture debate. In this chapter, we look at models favoring a reductionist approach to the issue, that is, models that attempt to examine the influence of variables independently of each other. In the next chapter, we look at models favoring a systems approach, that is, models that see variables in constant interaction with each other. The reductionist approach is the classic debate; the systems approach, the new debate.

The classic debate is just that, a classic. It seems to have been going on forever and, given the elements of the debate, is likely to continue going on forever. The players change periodically as do the particular statistics, but the fundamental question is always the same: What percentage of the variance is caused by nature and what percentage of the variance is caused by nurture? The fact that there seems to be no end in sight to the classic debate is perhaps one reason why many have abandoned reductionist approaches to development and have moved on to the more integrative, systemic ones.

Neither side in the classic debate claims ownership of all the variance, but both nativists and empiricists each claim ownership of most of it. The nativist position is reflected in a behavior genetic perspective but the empiricist position, as noted in the last chapter, is a bit harder to pin down. The fact of the matter is that there really is no particular empiricist position in the same sense that there is a behavior genetic position or, as we see in the next chapter, an evolutionary psychological position or a developmental systems theory position. In spite of the fact that (or perhaps because of it) the largest percentage of the research done within psychology and related disciplines published in journals such as *Child Development* or *Developmental Psychology* takes a strong empiricist perspective, there simply is no generally agreed-on theoretical argument as to how nurture operates. It is as if empiricists have simply accepted as a given that nurture affects development and now the question is to work out the details. I am not suggesting that the lack of a commonly shared, specific environmental perspective means that nurture does not affect development, because clearly it must; rather, it is only to say that the mechanism through which it does so may be less clear than we would like to think.

The Behavior Genetic Perspective

The modern behavior genetic perspective probably starts with Jensen's 1969 article, but in fact the questions now being asked by behavior geneticists are essentially the same as those of Galton. The methodology and analysis techniques have improved; the arguments and conclusions are now more data driven than what was once seen by Galton and his peers as self-evident conclusions, but the basic issues are the same, that is, what role does genetics play in defining the course of human development and, of actually more interest to behavior geneticists, what role does genetics play in explaining the differences between each of us in terms of the various dimensions of development?

An interest in course and variability is not unique to behavior geneticists; it is the stuff of all theory and research in human development. What is particular about the behavior genetic perspective is that its argument that attempts to explain course and variability exclusively through environmental variables, what Pinker (2004) refers to as the "blank slate" approach, is simply wrong. Instead, Pinker and others (Scarr 1996; Plomin and Asbury 2005; Plomin 2009) argue that much of this pattern and variability reflects genetics and, because genes and environment covary, much of the presumed environmental contribution to development is also to a large degree a reflection of genetics.

What is the logic and evidence to support such a claim? To behavior geneticists such as Bouchard (2009), there is quite a bit, clearly more than enough, according to them, to put the blank slate away for good. But to understand the arguments and the data, we first need to review some methodological and statistical points made in earlier chapters. First method and then statistics.

The bottom line of a reductionist approach to science is that variables that influence one or more behaviors can be teased apart so that the relative influence of each can be determined. How then can you do this with respect to nature and nurture? If you do nonhuman research, you simply engage in selective breeding so that you are able to produce stains of very inbred mice or some other species. Equally, you can then expose these highly inbred strains to very particular environmental manipulations and see what then happens. In effect, by controlling both nature and nurture, it presumably becomes possible, given a reductionist perspective, to see what causes what. Human researchers don't have it so easy. Selective breeding is fortunately out of the question, as are deliberate attempts to control one's environment. The human researcher is left with finding naturally occurring events that potentially can lend themselves to the partitioning of nature and nurture. As discussed earlier, these events are kinship studies and adoption studies.

Kinship studies compare the degree of behavioral similarity between individuals of differing degree of genetic relatedness. If, for example, the degree of variability between a sample of sets of monozygotic (MZ) twins is less than that of a sample of sets of dizygotic (DZ) twins, then, presumably, there is *prima facie* evidence for a genetic influence because identical twins share all their genes whereas fraternal twins share only half their genes. And if, in turn, the sample of fraternal twins shows less variability than a sample of cousins who, in turn, show less variability than a sample of individuals chosen at random, so much the better. As we will soon see, much research follows this logic, but there is a problem with it, one readily acknowledged by behavior geneticists: Twins share environments as well as genes. What is to say that the greater similarity among the identical twins isn't due to the fact that because they are identical, they are treated identically, or at least more similarly than fraternal twins would be?

For those particularly interested in kinship studies, there are a couple of potential, if not particularly common, ways around the problem (Reiss 1993). One is to find instances of MZ and DZ twins separated at birth and then see if the differences in the relative variabilities continue to hold. The second is to find instances in which DZ twins are mislabeled as MZ twins. Such things happen, especially if blood types are the same. If DZ twins are then raised as if they were MZ twins and still showed greater variability, then again the argument is that the differences in the degree of variability between the samples is genetic in origin. Although these two strategies do seem to eliminate the confounding of nature and nurture, the much more common way of doing so is through research on the more typical practice of adoption.

Adoption studies would seem to solve the confound because the adoptee is genetically unrelated to the adoptive parents. Therefore any degree of similarity would presumably reflect the environment and any degree of difference genetics. On the one hand, for example, if adopted siblings are more like the biological children of their adopted parents than a comparable group of children chosen at random, this would argue for an environmental effect. But if, on the other hand, adopted children were more like their biological parents than their adopted parents – even when separated from their biological parents at birth – then this would be evidence for a strong genetic effect.

Even here the partition is not perfect. Even if separated from her child at birth, the biological mother served as the prenatal environment for the then adopted child. One would really have to find a sample of children who had surrogate mothers prenatally and then were adopted at birth by still another family, fortunately a very unlikely circumstance. And then there is the issue of what adoptive parents are typically like, that is, they are not a very heterogeneous group, quite the opposite in fact. Potential adoptive parents go through a very thorough and rigorous screening process with respect to a variety of factors determining family stability and quality. Those approved are clearly "solid citizens." The more uniform the adoptive families, the less differential child-rearing influence they are likely to exert on their adopted children, certainly less than would be the case in a random sample of families. The less differential influence, the less measured the effect of the environment. The less the measured effect of the environment, the more measured the effect of genetics in reductionist models because if it isn't one, it's the other.

It is again important to note something mentioned in earlier chapters. Behavior geneticists are interested in differential influence, not influence per se. No one is suggesting that parents do not have an influence on their children with respect to what these children are like. Rather, the issue for the behavior geneticist is not what the children are like but rather how alike they are. Statistically, what the children are like is typically measured by mean differences: Do children reared one way do better in school, on average, than those reared a different way? But for the behavior geneticists, the issue is not average differences but rather is a comparison of differences in the degree of variability between groups of children. It might sound like the two are really the same but they are not, and, in fact, statistically, they are actually independent of each other. For example, there are data documenting the fact that adopted children's IQ scores correlate higher with those of their biological parents than with those of their adopted parents (i.e., the adopted child with the highest IQ score had a biological mother with the highest IQ score, etc.) but, at the same time, the adopted children's actual IQ scores are more similar to those of their adopted parents than to those of their biological parents, that is, the adopted children's mean IQ scores are closer to those of the adopted parents than to those of the biological parents.

Good research designs allow for the collection of data in the most scientifically rigorous way as possible. However, because data are almost always collected from a sample of a population rather than from the entire population, there still needs to be a way of determining the likelihood of the data from the sample being representative of the population from which they were drawn. Here is where the statistics come in and, in the case of behavior genetics, much of the controversy. Here is the equation first introduced in the last chapter.

$$V_{BP} = V_{GA} + V_{GNA} + V_{ES} + V_{ENS} + V_{GEcorr} + V_{GEint} + V_{ERR}.$$

It reflects the conceptual model used by behavior geneticists to identify possible sources of variance affecting V_{BP}, the variance in the behavioral phenotype of interest to the researcher, that is, the dependent variable. More complicated research designs might involve more than one dependent variable, but the possible independent variables would still be the same. V_{GA} reflects the degree of variance accounted for by additive genetic effects. Additive genetic effects are seen as exerting a quantitative influence on some trait and should remind you of the discussion of Mendel and his peas from your high school biology class. V_{GNA} refers to nonadditive genetic effects and reflects the fact that genes often interact with other genes as they exert their influence of development. Typically, behavior geneticists treat the two as one variable because their concern is not with the mode of action of specific genes but rather with their ultimate expression. V_{ES} and V_{ENS} refer to shared and nonshared environmental variance, respectively. Variance that is shared is assumed to exert the same influence on all those receiving it, as might be predicted to be the case in a study of parental child-rearing strategies. Nonshared environmental variance refers to the particular environmental experiences that are unique to each of us. The relative importance of these two sources of environmental variance is a hot issue because the behavior geneticists argue that there is very little evidence for the influence of shared environmental variance. Shared environmental variance is, however, the "bread and butter" of the environmentalists and, as such, they don't take kindly to any claim that says that what they study is of little significance. V_{GEcorr} refers to the correlation that is seen as existing between genetic and environmental factors. Most typically, behavior geneticists argue, the two work in concert. For example, we might expect tall parents to be more likely to hang a basketball net over the garage than short ones. V_{GEint} refers to the likelihood of an interaction between nature and nurture. Unlike the case of a gene–environment (GE) correlation, in which the two variables are seen as typically working in the same direction, there is no similar assumption made about interactions. The status of GE interactions is also a hot topic, as the behavior geneticists say that there is little evidence for them in human research and the system theorists argue that GE interactions are the paramount element in defining the course of development. Finally, V_{ERR} is the error term, the recognition that no matter how good a research design, it probably doesn't capture everything or control for everything.

As I discussed in Chapter 3, each of the terms in this equation is highly controversial, especially those dealing with interactions and environmental variance. It isn't a matter of the mathematics per se; rather, it is a matter of how valid the statistical representation of the behavior of interest is. If the representation is incorrect, it really does not matter how accurate the math is; as the computer folks like to say, garbage in, garbage out. An example should help make this all clearer, one dealing with what is certainly the most controversial topic in the behavior genetic literature – race. It concerns an article by David Rowe (2005), "Under the skin: On the impartial treatment of genetic and racial differences" and a rejoinder by Richard Cooper (2005).

Rowe's argument is pretty straightforward. If individuals differ in terms of physical characteristics and in terms of susceptibility to various medical conditions and if these differences have a genetic base, and, further, if these individuals can be categorized in terms of distinct racial groups, then why should we not expect similar patterns of difference in terms of measures such as intelligence? He argues that such a hypothesis explains why four fifths of the black population fall below the white population mean of 100 and "why Black individuals are underrepresented in earned doctoral degrees in the natural sciences and mathematics" (2005, p. 62). He then goes on to argue that these racial differences evolved over long periods of time as populations became reproductively isolated from each other over the course of human evolution. Because each population was presumably exposed to different conditions of natural selection, each evolved specific traits and abilities in response to these selective pressures. He then reports data showing such race-based differences on measures of intelligence and even takes the argument further by also providing data showing that "the IQ of mixed-race children fell between those of the homogeneous Black and White children, as expected under a genetic hypothesis" (p. 67). In further support of this "in-between hypothesis," Rowe notes a correlation between skin color and IQ; in particular, for blacks, the lighter the skin color (presumably a reflection of the percentage of "white genes" in the individual), the higher the IQ. Rowe then dismisses the possibility that such a correlation could reflect greater racial discrimination against dark-skinned compared with light-skinned blacks: "Yet no mechanism for this discrimination effect has been proposed that is viable. In the United States, Jews and Asians have both endured significant discrimination but without apparent harm to their IQs" (pp. 67–8). He concludes by arguing that the role of science is to identify differences with impartiality and to put politics and ideology aside when doing so. He does not discuss the public policy implications of such racial differences, if, in fact, they were to be accepted as valid.

Cooper (2005) gets right to the point in his rebuttal:

> The technical errors contained in Rowe's (2005) article include misuse of broad scientific concepts and incorrect or biased misinterpretation of specific scientific data. The author's broad argument assumes that a quantity definable as "intelligence" exists (in contradistinction to the view that multiple types of cognitive functioning can be identified that are valued and manifested differently, conditional on the setting of the observer), that intelligence can be measured with "IQ tests," that demographic groups known as "continental races" divide humans into discrete categories on the basis of important concordant variation in genetically determined traits, that molecular genetics can (or will) make it possible to define the architecture of complex traits in terms of "genes for X or Y" (i.e., "genes for intelligence"), and that the significant variation in polymorphisms in those genes overlays with the traditional demographic categories, such as those promulgated by the U.S. government. (p. 71)

Cooper then goes on to argue that racial categories are much more a social than biological reality, that the argument for IQ differences being a reflection of reproductive isolation between groups "is so improbable that it can be discarded" (p. 73), and that social determinism, especially with respect to the relation of skin color to IQ, is a much more plausible explanation for test score patterns within the United States. With respect to Rowe's argument that Jews and Asians have been discriminated against without any apparent effect on their IQs, Cooper argues that "this blindness to the historical patterns of cultural assimilation and anti-Black discrimination in the United States and its influence on socially determined traits voids any claim the author might make to integrity and rigor in examining this question" (pp. 74–5).

Whatever one might think about the merits of Rowe's hypotheses or Cooper's rebuttal, the point here is that Rowe's arguments are based on the specific statistical model described above, the model that serves as the point of reference for all behavior genetic research. This does not mean that behavior genetic research is racist, but it does mean that attempts to partition genetic and environmental variances within a reductionist paradigm almost inevitably lead to questions about the validity of the evidence for such alleged differences, the hypothesized causes of such alleged differences, and the public policy implications if the findings are, in fact, valid.

Although there are many people pursuing behavior genetic research, the work of two of these scholars is perhaps most significant, both in terms of the volume of their efforts and in terms of its visibility. These two individuals are Robert Plomin and Sandra Scarr. It is worth looking closely at the work of each because doing so provides an excellent illustration of what behavior genetic research is all about.

The Work of Robert Plomin: Over a period of many years, Plomin and his colleagues (Plomin and DeFries 1983; Plomin and Daniels 1987; Plomin et al. 1993, 1994a, 1994b; Plomin and Asbury 2005; Oliver and Plomin 2007; Plomin et al. 2007; Plomin 2009) have identified a number of examples of behavioral variability correlating with genetics, typically defined through kinship and adoption studies. More recently, he has begun the process of attempting to identify the particular genetic markers that might be associated with behavioral variability on one or more dimensions.

Plomin's early work served to document the importance of genetics on behavior variability. Employing both kinship and adoption studies involving a variety of measures across a variety of behavioral domains, Plomin (1990a) consistently found that often up to 50% of the variance on any particular behavioral measure of personality or cognition or attitude or vocational interests or delinquency or learning difficulties could be accounted for by additive and nonadditive genetic factors:

> To appreciate the power of heredity, it should be noted that in the behavioral sciences it is rare to explain this much variance for any behavior. Genetic influence is so ubiquitous and pervasive in behavior that a shift in emphasis is warranted: ask not what is heritable; ask instead what is not heritable. So far the only domain that shows little or no genetic influence involves beliefs such as religiosity and political values; another possibility is creativity independent of IQ. (Plomin 1990a, p. 112)

What then of the remaining variance? Once genetic variance is removed, what you are left with is shared and nonshared environmental variances and GE covariance and interaction plus error variance. How is the remaining 50% apportioned among these four variables and the error term? A reasonable presumption would be that if there is strong evidence that genes have a direct influence on behavior, then so too should the environment. As such, we should expect to see significant correlations between, for example, parenting styles and child outcomes, particularly among siblings who jointly experience their parent's child-rearing approach. This is certainly the argument that environmentalists make in their research and in the reporting of their data. But, according to Plomin, this actually is not the case at all. In the first place, he argues that the amount of variance accounted for by shared environmental factors, even when it is reported by environmentalists, is actually quite small and only looks significant at all because most environmental research looks at only the interactions of parents with an individual child rather than at parents in dealing with all their children.

For Plomin, the problem of shared environmental variance is one of method. Environmentalists, he argues, identify a sample of children from different families and then determine the degree of relationship between some aspect of the child's behavior and some aspect of the parent's behavior. However, because only one child per family is typically involved in such research (often simply because the researcher is primarily interested in children of a particular chronological age), there is, in fact, really no way to see how significant either parent's behavior is on the child's development because with only one child per family sampled, there is no way to measure variability within that family. In other words, what if it turned out that even when parents do the same thing in dealing with their different children, the children nevertheless turn out quite differently from each other? According to Plomin, this is, in fact, exactly what happens in families, that is, shared environmental variance is a negligible influence and nonshared environmental variance is a significant influence.

In his 1987 review paper with Daniels (Plomin and Daniels 1987), Plomin documents the evidence for such an argument and then offers an explanation for the question "Why are children in the same family so different from one another?" They note that in studies of personality measures, as much as 80% of the reported environmental variance is due to nonshared environmental variance and that this is the case even when one sibling exhibits a significant psychological disability. With respect to measures of cognition, they do report that shared environmental variance does appear to have as significant effect, but there is caveat attached to the finding; the correlation holds only for young children, decreases during middle childhood, and becomes negligible by early adolescence:

In summary, nonshared environmental influence is a major component of variance for personality, psychopathology, and IQ (after childhood). We conclude that nonshared environment explains perhaps as much as 40% to 60% of the total variance for these domains. Although one can quibble with the magnitude of our estimates, they would have to be substantially in error before they would affect our argument that most of the environmental variance is nonshared. (p. 6)

Why? Why do families seem to have so little shared or common influence on their children, at least as measured by variability estimates? Note that the question is not whether parents influence their children; rather, it is why this influence seems so specific to each child in that family. First, unsystematic events such as accidents or trauma, divorce, or economic recessions may affect some but not all children in a family or affect siblings in different ways, especially if there are wide age gaps between the siblings. Second, even when parents report that they see themselves as treating their children in the same manner, the children's perception is less so. Plomin and Daniels note, for example, that whereas parents' agreement on how similar they treat their children ranges, in different studies from 0.38 to 0.65, their children see things differently: Their perception of similar parental treatment averages only 0.20. These data do not necessarily mean that the parents are not actually treating the children similarly in an objective manner, only that the children's perceptions of the parents' efforts differ. Third, parents are not the only influence on children's development. Children interact with their siblings and with their peers, they find themselves in different classrooms, and they engage in different out of school activities. Each of these is unique to each child.

So why do parents, to a significant degree, treat their children differently? The answer, for Plomin and Daniels, is obvious. Parents treat their children differently because the children are different and the reason the children are different is because their children differ genetically. Therefore, MZ twins present a more similar behavioral profile than DZ twins or siblings because of their degree of genetic similarity, and adopted children differ so much from the biological children of their adopted parents because they have little, if any, shared genetics. Said another way, the nonshared environmental data say that children have more of an influence on their parents' behavior than parents do on their children's behavior. Plomin and Daniels argue that this "direction of effects" issue, first raised some years ago by Bell (1968), is an essential piece of the puzzle in trying to understand why there is so much variability between children within the same family. In other words, for behavior geneticists, the environment is not the independent variable, that is, the cause, but rather the dependent variable, that is, the effect. Environments don't cause children's behavior; children's behavior causes their environments.

Plomin's more recent work has served to provide additional support for his behavior genetic arguments and, in addition, to begin to move beyond the simple demonstration of correlations between behavioral variability and degree of genetic relatedness to now look for the specific molecular genetic clusters responsible for such associations. Much of this work involves a very large-scale longitudinal study know as the Twins' Early Development Study (TEDS) (Kovas et al. 2007; Oliver and Plomin 2007). The TEDS study involves data collected from more than 12,000 children in the United Kingdom. Children were sampled at the ages 2, 3, 4, 7, 9, and 12 years on a variety of measures related to a broad range of developmental dimensions, including learning abilities and disabilities. Data included measures of family structure and function, a variety of prenatal and postnatal biological measures, teacher assessments of children's development, and assessments of children's cognitive development, including measures of reading, language development, and mathematical ability.

In support of earlier research, Plomin and his colleagues report that evidence of shared environmental influence is highest when children are infants and preschool age and then wanes as children move through middle childhood into adolescence. Moving beyond earlier findings, the scale of the TEDS data, especially the repeated ability measures taken across a variety of verbal and nonverbal cognitive domains, allowed them to gain a better sense of how genetics influence development. Two findings are seen as particularly noteworthy. First, the similarity of twin variability patterns across the different measures leads Plomin to argue for the essential role of *generalist genes* in the regulation of children's cognitive development (Butcher et al. 2006). In other words, the fact that children who do well in, for example, reading also do well in mathematics leads Plomin to argue that the set of genes influencing ability in the first domain is the same as influencing ability in the second:

> Multivariate genetic research on cognitive abilities consistently reveals that genetic correlations among diverse abilities are typically about .80. This suggests that mainly the same set of genes affects cognitive abilities as diverse as memory and spatial in addition to information-processing measures. Moreover, there is some evidence that the genetic overlap among cognitive abilities increases during development, with genetic correlations reaching 1.0 in late adulthood. (pp. 145–6)

These generalist genes are seen as functioning in a quantitative manner. Many genes, working in concert, each contribute a small influence to the total variability. This generalist pattern is described in contrast to those rare but often fatal monogenetic influences in which the presence of one particular gene does determine developmental outcome. One significant implication of this generalist pattern is that efforts to look for the "gene for" some developmental measure are likely to be futile because no one gene alone is seen as exerting a significant influence on the outcome measures. Rather, genes are *pleitropic*; they each have multiple effects and are likely to each be active in many regions of the brain.[1]

[1] It is worth noting that this generalist argument stands in sharp contrast to the evolutionary psychological position discussed in the next chapter. For the evolutionists, the brain is not a general but a highly modularized organ. It is also worth noting that whereas Plomin finds little if any evidence for sex differences in the TEDS data, the evolutionists see sex differences in virtually all aspects of our behavior. Finally, it is also worth noting that the developmental systems theorists also discussed in the next chapter would take serious exception to Plomin's argument that

Plomin also believes that the generalist gene hypothesis leads to the conclusion that children with either learning disorders or exceptional learning abilities do not have unique genotypes compared with more typical children but rather are simply children at either end of the generalist continuum. As he puts it, the abnormal is normal.

As is true of virtually all human behavior genetic data, the TEDS data are correlational; they show a pattern of relationship between variables. Plomin believes, however, that it is reasonable to argue cause from such correlational data when those data involve genetics because, he argues, the "correlations between DNA differences and behavioral differences can only be explained causally in one direction: DNA differences cause behavioral differences" (Butcher et al. 2006).

Although Plomin's work, unlike Rowe's, does not deal with highly controversial issues such as race, it nevertheless raises the same questions about how best to understand the implications of such data in democratic societies. Plomin certainly recognizes this concern (Plomin and Asbury 2005), in particular the concern that behavioral genetic research raises fundamental questions about the meaning of equality. Plomin's response to such concerns is direct. He argues that we all deserve legal equality regardless of our differences, whatever the source. But he is also quick to point out that to deny that some potion of human diversity is genetic in origin is simply choosing to "stick our heads in the sand":

> The basic message of behavior genetics is that each of us is an individual. Recognition of, and respect for, individual differences is essential to the ethic of individual worth. Proper attention to individual needs, including provision of the environmental circumstances that will optimize the development of each person, is a utopian ideal and no more attainable than other utopias. Nevertheless, we can approach this ideal more closely if we recognize, rather than ignore, individuality. (p. 96)

The Work of Sandra Scarr: Unlike Plomin, Sandra Scarr has not been reluctant to talk about the social implications of a behavior genetic perspective. Rather, she has loudly and frequently critiqued (Scarr 2009) what she sees as the "intellectual bankruptcy of socialization research." Over a long and active academic career, using the same kinship and adoption research models as other behavior geneticists, she has advanced a number of arguments concerning the impact of parents on children's development, on matters of race, on socioeconomic influences on development, and on the changing influence

gene–behavior correlations can be treated causally because they function in only one direction. Systems theorists would argue quite the opposite.

of genetics on development across the life span (Scarr and McCartney 1983; Scarr and Weinberg 1983; Scarr 1991; Scarr and Ricciuti 1991; Scarr 1992, 1993; Scarr and Weinberg 1994; Scarr 1996, 1997). I made reference in Chapter 1 to Scarr's 1991 Presidential Address to the Society for Research in Child Development as perhaps the point where her arguments first created such a flashpoint. This was not, however, the first time that Scarr had talked about the role of genetics in development but, as was true of Jensen's 1969 paper in the *Harvard Educational Review*, her decision to do so in the midst of a sea of socialization researchers surely heightened the impact of her comments. So what is she saying?

> The most general principle is that *people are not randomly assorted in their environments* [italics in original]. Social science research finds ubiquitous correlations between people and the environments in which they are found. The typical, usually unspoken, assumption is that some people are just unlucky and wake up one day to find themselves poor and ill-educated, whereas others are just lucky to have had opportunities for higher education and well-paying jobs. The unspoken implication of most "environmentally" motivated research is that merely providing have-nots with the same opportunities as the haves will equalize their outcomes. I will show that this assumption leads to false predictions about interventions to improve the lot of the have-nots, because people are not randomly thrust into their environments; rather, to a large extent, they make their own environments by expressions of genetic variability. This is not to say that morally or ethically people deserve their fate – only that their personal characteristics are correlated with them. (Scarr 1996, p. 205)

There are many aspects to Scarr's work, but with respect to the nature side of the classic debate, three are perhaps most germane. The first concerns her stinging critiques of what she refers to as "socialization theory," that is, her view of the nurture side of the classic debate. The second is her perspective on the relative influence of nature and nurture across the life span. The third is her take on the interplay between socioeconomic status (SES) and human development.

For Scarr, socialization theory is concerned with what a behavior geneticist would refer to as shared environmental variance. And if you read the nurture-oriented socialization literature, you find there are many reports showing significant associations between some aspect of a child's context, often his or her parenting experience, and some child outcome measure such as intelligence or some measure of personality. But the behavior genetic literature research shows virtually no variance being explained by shared environmental variance. So how can the nurture folks find significant associations when the nature folks claim there is nothing there to be found?

One answer, of course, is that they are using different dependent measures: variability for the nativists and group mean differences for the empiricists. But, according to Scarr (1997), there is a more fundamental reason for the discrepancy. Because most socialization research looks at only one child per family, there is, according to her, a total confound between the effects of genetics and the effect of environment, and as such there is actually no way for the socialization theorists to demonstrate that their correlations between, for example, parent behavior and child behavior, actually reflect environment rather than genetics. And, Scarr argues further, because kinship and adoption research designs do allow the disentangling of genetic and environmental variables and do show genetics to be a much more significant cause of variability than environment, then it is most reasonable to conclude that the associations reported by the socialization theorists are more likely a reflection of the shared genetics of parents and their children than their shared environments. In other words, as far as Scarr is concerned, all the socialization literature is doing is providing more evidence for the role of genetics in directing children's development.

For Scarr, GE confounding, or, more correctly, GE correlations occur across the entire life span but the magnitude of the correlations change as children grow older. Scarr identifies three types of correlation associated with age: *passive*, *evocative*, and *active*. Passive correlations most typically describe GE correlations early in life, evocative correlations become more prominent as children move from infancy into the preschool years and beyond, and active correlations most typically describe GE interactions as children move through middle childhood into adolescence and then the adult years. Scarr (1996) sees the relative influence of the three correlation patterns explaining why evidence for shared environmental variance is highest when children are young and then decreases over the childhood years so that, by adolescence, to behavior geneticists, shared environmental variance accounts for virtually nothing.

Passive correlations reflect the GE confound discussed above, namely, parents share both genes and environments with their children. So one should not be surprised that parents who would be seen as having a genetically based musical aptitude are likely to have a lot of musical instruments and other musically related material around the house. Their child therefore likely inherits not only the parent's genetic musical competence but a rich music environment as well. By the same token, a parent with significant reading difficulties is less likely to have books around the house but is likely to have a child who also demonstrates reading difficulties, both because of inheriting the parent's reading problem and because of growing up in a literacy-poor environment.

The correlation is described as passive simply because, for better or worse, the child has nothing to say about it.

Children are different from each other and as such evoke different reactions in others. These reactions serve to positively or negatively reinforce the child's expressed behavior, which in turn serves to help shape a child's sense of self. Because, according to Scarr, these behavioral differences have a significant genetic base, the child's genotype is seen as evoking or directing the child's environment, that is, the child is seen as making his or her own environment. Such evocative correlations are seen as explaining much of both parent and teacher reactions toward children as well as accounting for much of the fate of adults in such matters as dating and mating and job success.[2]

Sooner or later we each grow up and are able to seek, to one degree or another, our own way in the world. To the extent that we are able to chart our own course, Scarr argues that we tend to gravitate to settings and situations that we see as a good fit for our interests, talents, and personalities. Those who love reading gravitate to the library, those who love sports to the athletic field, and, in both cases, find other like-minded individuals to become friends, colleagues, and partners. As such, as we move into the adult years, we become increasing active in intentionally directing the course of our own development. And this intentional active life structuring is a reflection of the increasing expression of the genotype, now "freed" from the shared environment.

Scarr's concept of the "good enough parent" logically follows from her three correlation patterns. Because the significance of the parents' genetic contribution to their children's development increases (i.e., nonshared environmental variance increases) and because the significance of parent's efforts to influence their children through their environmental actions decreases (i.e., shared environmental variance decreases), Scarr argues that her research helps parents relax a bit in terms of ensuring their children's developmental outcomes, that they need to be only "good enough":

Good enough, ordinary parents probably have the same effect on their children's development as culturally defined super-parents. This gives parents a lot more freedom to care for their children in ways they find comfortable for them, and it gives them more freedom from guilt when they deviate (within the normal

[2] Scarr's discussion of evocative correlations looks at the matter only from the perspective of the child, that is, the world is seen as simply responding to the image the child presents to it. But the situation is surely more complicated because not only would, for example, a child's behavior evoke responses from the teacher but, equally, the teacher's behavior would evoke responses from the child. The process is likely more complicated than Scarr depicts it, especially with respect to disentangling genetic and environmental sources of variance.

range) from culturally prescribed norms about parenting. . . . If I were to advise parents, I would say that the research supports the idea that parents need to provide *opportunities, not prescriptions* [italics in original] for their children. A rich and varied environment of opportunities afforded by the family will provide every child with the possibility of becoming the best he or she can be. For some parents with the fixed ideas about their offspring's outcomes, this will be anxiety-provoking advice, but given that nearly all parents want their children to grow up to be well-functioning and happy adults, this is good advice, based on behavior genetic research. (Scarr 1996, p. 221)

But what if parents are unable to provide their children with a rich and varied environment? What happens then? Or, to put it more directly, what role does SES play in defining developmental outcome and, if it does play a role, why? The question as to whether SES plays a role is pretty much answered: It does. There are all sorts of evidence from both nativist and empiricist perspectives attesting to the fact that children from low-income households generally do not do as well on measures of intelligence or achievement and that they typically finish fewer years of education, are more likely to end up in low-paying jobs, and are more likely to experience difficulties across their adult years than children from economically advantaged backgrounds. But here the agreement ends. Scarr argues that socialization theorists claim that these SES differences are largely a reflection of a degree of opportunity that in turn reflects social structure. Improve schools, improve housing, improve access to health and social services and the achievement gaps between children at different SES levels will disappear. The fact that, in spite of efforts to do just this, these differences have not yet disappeared is seen, according to socialization theorists, as reflecting the fact that we simply have not tried hard enough yet: Remember Hunt's response to the 1969 Jensen article claiming that compensatory education has failed was "Has compensatory education failed? Has it been attempted?"

Scarr sees things differently. Consistent with her take on parenting, Scarr sees SES differences not as a reflection of opportunity but rather as a reflection of genetics. Intellectual differences between children of different SESs is a reflection of the genes they have inherited from their parents. Further, because mobility is possible between socioeconomic levels, those whose disadvantage might in fact be due to circumstance rather than intelligence are the ones most likely to either move up the ladder and/or provide a set of genes that increases the likelihood of their children doing so. If this is the case, the social policy implications are profound. Opportunity makes it possible for those who can benefit from such an opportunity to create a better life for themselves and their offspring; those who do not have the capacity to make use of this opportunity are left behind, that is, rather than equal opportunity eliminating SES differences, it actually, according to Scarr, can serve to increase SES differences.[3] Again, it is nonshared environmental variance rather than shared environmental variance that is the defining element.

Scarr does make one exception to this generalization, however (Scarr and McCartney 1983; McCartney and Berry 2009). Shared environmental variance does account for a significant degree of variability in situations in which children are abused or neglected or in which the degree of poverty is severe. This "nonlinear model" argues that, on the one hand, below a certain threshold, the environment is so limiting in the opportunities and supports it is able to provide the child that irrespective of the child's genotype, adequate development is not possible. On the other hand, children living above this threshold do have a reasonably adequate environment to support development, and therefore, for these children, genetics does become the more significant variable. This notion of threshold is used by Scarr to explain why children from very disadvantaged settings show significantly greater benefit from high-quality early intervention programs than children living above the threshold. For children above the threshold, the environment is already seen as good enough, and, as such, even more enrichment is not seen as having any additional benefit, but just the opposite is true for children below the threshold.

Some Tentative Conclusions About Nature

The nativist position reflected in behavior genetic research offers a great deal of evidence attesting to the role of genetics in human development. The research is reductionist in perspective, is highly statistical in its methodology, and rests on a number of controversial assumptions, such as the equal-environments assumption or the representativeness of located twin samples. But this current body of work is not as easily dismissed as the writing of those such as Galton and Fisher, who already "knew" that there were genetic differences between groups and then set out to document them. Since Jensen's 1969 paper, the consistent findings about correlations between cognitive and personality measures and degree of genetic relatedness, about the developmental status of adoptees, and even about the differential effect of early intervention programs force us to recognize that something is going on here. It may not ultimately be found to be as the behavior geneticists describe or explain it, but it clearly means that any effort to explain probably any aspect of development without a serious understanding of the role and place of genetics is doomed to fail.

[3] I return to this issue in Chapter 6 when I discuss the controversial work of Herrnstein.

The problem of course is to understand what that role and place are, especially in terms of differences between individuals and groups and in terms of the issue of the limits on development. These are difficult questions to ask, much less to answer, especially in a democracy that places so much emphasis on life, liberty, and the pursuit of happiness. To even suggest that some, especially some groups, might be more successful in these pursuits than others, no matter what the opportunity available, is political dynamite. However, to put the claims of Jensen, Rowe, and others to rest will require a much deeper understanding of genetics than we now have and will certainly require much more sophisticated methodology than is now used. But that doesn't mean that we don't look for these better ways.

The Environmental Perspective

One would think that because most behavior genetic research is designed to show that nature accounts for more variance than nurture, environmental research would attempt to show just the opposite. A reasonable assumption, but typically not the case. There certainly are studies that respond to behavior genetic claims, but these have typically been in response to specific papers, as was the case with Jensen's 1969 paper (e.g., Hunt 1969; Kagan 1969) or in response to Scarr's 1991 SRCD Presidential Address paper (Scarr 1992), or in response to articles by Plomin (e.g., Richardson and Norgate 2005, 2006). But, for the most part, pick up almost any mainstream child development journal and you get the sense that most, if not all, of the authors have not even heard about behavior genetics. Well, admittedly that is an exaggeration; environmentalists have certainly read the work of the behavior geneticists but it doesn't seem to have made much of an impression on their own work.

Why? If behavior genetic research is primarily designed to show that nature is more important than nurture, why isn't environmental research designed to show that nurture is more important that nature? There are probably several reasons, none of which includes the belief that environmentalists do not think nurture *is* more important than nature. Rather, as is the case for all reductionist research, the issues are primarily methodological. First, because environmentalists are trying to document the effect of efficient causes (i.e., those external to the organism), the focus is on finding mean differences between groups receiving different antecedent treatments. As mentioned in the previous chapter, for environmentalists, the goal is to reduce variance because the smaller the within-group variance, the more powerful the effect of the antecedent or independent variable, that is, the more variance it accounts for. For behavior geneticists, just the opposite is true. Because in behavior genetic research involving humans at least, antecedents cannot be controlled, the behavior geneticists look for correlates of behavior with, typically, adoption or kinship status and to document these associations; the more variability, the better.

Second, sample composition is very different in the two approaches. Because of the focus on variability, behavior genetic studies tend to have larger samples. Environmental research typically has smaller sample sizes. The reason for the difference reflects the nature of the research. Behavior geneticists typically gather information about shared environmental variance either through self-reports or parental reports or paper-and-pencil-type evaluations; environmentalists are more likely to observe over some period of one or more times, as might be the case in studying children's play patterns, or work directly with children in a one-on-one lab setting, as might be the case on seeing how different materials or instructions affect some aspect of cognitive development. And for some behavior genetic research, shared environmental variance is simply defined by what variance remains after all other factors are partitioned out. In this case, no environmental data are actually collected.

Third, behavior genetics samples are very specific and tend to have more than one child or adult per family; environmental samples are more likely to be random and rarely involve more than one person per family or one child per family in the case of research on parent–child relationships. For environmentalists, the specificity of the behavior genetic sample raises questions of generalizability, and, for the behavior geneticists, the environmentalist's sample is seen as confounded because, according to them, there is no way to know if the data reflect nature or nurture. As I have mentioned before, such differences separated the two groups 100 years ago, they continue to do so today, and they will probably do so 100 years from now unless our models become truly interactive and systemic rather than remain reductionist. Said another way, it isn't that environmentalists do not recognize the fact that genetics must play some role in development: rather, it seems that they are concerned with only the role that environment plays. And this focus on the environment is no doubt what has led Pinker (2002) to argue that environmental research favors a *blank slate* model.

Unlike the behavior genetic position (as well as the two discussed in the next chapter), there really is no "official" environmental position on the nature–nurture debate other than a shared belief that nurture is more important than nature. And, by extension, there really is no one person whose work defines an environmental position. This situation probably reflects two factors. First, the environment is a pretty big place and to try to get a handle on it is a very challenging task. Most environmentalists rather seem content to each explore the effects of some limited set of antecedents on some limited set of dependent variables. Trying to then put all of these focused efforts together to get the really big picture has so far proved difficult. Perhaps one reason why integration has proven so difficult relates to the second factor, namely, that there really is no such thing as an environmental unit of measurement, a point made by Horowitz some years ago (1993)[4]:

> There is also no agreement as how to best parse or define the environment in which a person is growing up. There are no standard units of environmental variables, and the definitions of developmentally facilitative environments find no consensus. At best, the efforts to manipulate environmental variables focus on gross dimensions; at worst, the efforts to define and manipulate the environment are laden with value judgments about good (positive/effective) and bad (negative/ineffective) that have minimal empirical support. (pps. 347–8)

There may not be a consensus as to an environmental unit of analysis, but this does not mean that people have not been trying to establish one. The next section of this chapter examines three such efforts to do just that, the work of Bronfenbrenner and Evans (1999; Bronfenbrenner and Evans 2000; Bronfenbrenner 2001), the work of Elder (1995, 1998), Elder et al. 1996, and the work of Evans and English on SES (Evans and English 2002; Evans 2004).

The Work of Urie Bronfenbrenner: Most people who know of Bronfenbrenner's work probably know it in terms of the four-tiered social ecosystem presented in his 1979 book (Bronfenbrenner 1979). As well known as this depiction of the environment is – virtually every child development text has an image of it – it is also important to recognize that the 1979 model was an early representation of the social ecosystem, one that underwent some significant changes after that date. And, as Tudge et al. (2009) point out, it is therefore essential to know which model an author is referring to when talking about Bronfenbrenner's work.

The 1979 model presented a four-tiered, nested model of the social environment. Our day-to-day lives exist at the level of the *microsystem*, the place where we actually interact with others in our lives and encounter the various

[4] It is important to point out that empiricists are not the only ones whose work is hampered by a lack of shared agreement as to a unit of analysis. Truth be told, the behavior geneticists are vulnerable to the same criticism. When, for example, a behavior geneticist reports a correlation between some dependent measure and some degree of genetic relatedness, we still don't know what in the genome is presumably responsible for the association, especially because efforts to show specific gene–outcome relationships have been rather spotty and, when found, hard to replicate. In fact, for many geneticists, even defining what a gene is has ironically become increasingly difficult as we continue to learn more about genetic and epigenetic mechanisms (Meaney, 2010). So, for those of you keeping score, in the category of "clearly defined unit of analysis," the score so far is zero to zero.

elements of our lives. Typically, microsystems involve a component of family, a workplace or school, peers, perhaps a religious setting, and so forth. The number and types of components are unique to each individual and would change over extended periods of time. Some individuals have many distinct elements to their microsystems; others, a simpler life. In either case, the defining element, again, is that the microsystem is where we live our lives.

Lives, especially complicated ones, take some degree of coordination in order to run smoothly. Bronfenbrenner's second tier, the level of the *mesosytem,* reflects this reality. How well do the demands of home and work mesh? How well do adolescent's peer group pressures mesh with family expectations? How able are adults to maintain some degree of intimacy and romance in their lives and, at the same time, see to the needs of three young children? The greater the degree of coordination, the more effective the environment is in supporting that individual's development. It goes without saying that some people's mesosystems work better than others and that the likelihood of a well-coordinated mesosystem is not a random event. Those with more power and privilege in society are better able to coordinate the various elements of their microsystems than those less fortunate. Class begins when the teacher arrives, not the student.

Many environmental events over which we have little if any control affect our lives. War, natural disasters, acts of Congress, economic downturns, all have significant impacts on our lives, but we have little influence on their occurrence. Nevertheless, the impact might be very significant, and events such as these are represented by Bronfenbrenner at the level of the *exosystem.* Imagine coming to work one morning only to find that your company has been sold, that the plant is closing, and that you have 30 minutes to gather your belongings and leave. And if the plant is the primary employer in a small town, imagine the impact on all the others businesses in the town as well as on the tax revenues that support public services.

The broadest level of the ecosystem is not a place per se but rather the set of beliefs and values that is held more or less collectively by members of a group. What does it really mean when the U.S. Constitution says that all citizens have the right to "life, liberty and the pursuit of happiness"? What does it mean that all citizens have "equal protection under the law"? Such matters and questions exist at the level of the *macrosystem,* and they in turn are typically interpreted and resolved at the level of the exosystem, most often through the courts or the legislative process. How intrusive should the government be in our lives? Does the government have the right to tell us what lightbulbs to buy, what we can and cannot do in the privacy of our own bedrooms, whether or not we should be required to purchase health insurance? What services should the government be expected to provide? Should government run the prisons and schools or are these responsibilities better left to the private sector? Who should we turn to for health care? The Constitution of the United States has nothing to say about lightbulbs or health care or matters related to intimacy and reproduction, but invariably these are matters that come before our courts and legislatures because they reflect efforts to intuit and interpret the values in those documents that define out macrosystem. And the decisions made by these courts and legislatures do affect the daily lives we lead within our microsystems and, by extension, how well these newly defined aspects of our lives fit together at the level of our mesosystems.

Following the 1979 book, Bronfenbrenner (1999, 2001) continued to refine the ecological model into a *bioecological* model:

> To turn, then, to the "new beginning" since 1979. The first important theoretical development after the publication of the original ecological model was the introduction of a critical distinction between *environment* and *process*. Traditionally, in developmental research, such phenomena as mother–child interaction – and, more generally, the behavior of others toward the developing person – have been treated under the more inclusive category of environment. In the *bioecological model*, a critical distinction is made between the concepts of environment and process, with the latter not only occupying a central position, but also being defined in terms of its functional relationship both to the environment and to the characteristics of the developing person. (1999, p. 4)

In effect, by making a clear distinction between environment and process, Bronfenbrenner was recognizing that the four levels in the 1979 model were best recognized as *social address* variables, that is, ones that speak more to the position of a person in a social system than to the person's interpersonal experiences within that system. Clearly the two are closely related, but this differentiation now makes clearer that there is more to understanding individual behavior and development than knowing, for example, that person's SES or marital status or work history. The revised bioecological model was defined in terms of a series of propositions.

The first and second propositions Bronfenbrenner described in a 2001 paper (Bronfenbrenner 2001) define development as a reciprocal process taking place between an "active, evolving biopsychological human being" and the elements of his or her immediate external environment. To be effective, this reciprocal interaction must occur on a "fairly regular basis over an extended period of time." These reciprocal interactions occurring in the immediate environment (i.e., the 1979 microsystem) are referred to as *proximal processes*.

In the third proposition, Bronfenbrenner notes that the form, power, and content of these proximal processes are functions of both of the developing person and of both the immediate and more remote environments (i.e., the 1979 exosystem and macrosystem), the nature of the expected developmental outcomes, and the degree of continuity and change taking place over time (i.e., the 1979 mesosystem). This conceptualization is then summarized as a *process–person–context–time* (PPCT) *model.*

In the remaining propositions, Bronfenbrenner lays out the specific elements that a developing person must experience in order for his or her development to be optimal. These experiences include a strong emotional attachment to those committed to the child's welfare that leads to the internalization of parental values. This progressively more complex reciprocity is aided by the presence of a *third person*, who "assists, encourages, spells off, give status to and expresses admiration and affection for the person caring for the child" (2001, p. 10). In such an arrangement, the child is seen as most likely to be exposed to and acquire the variety of social and cognitive skills necessary for full development. Bronfenbrenner, who had long been a critic of American society, was not particularly sanguine about such a development course happening for most American children:

> In the United States it is now possible for a youth, female as well as male, to graduate from high school, or even university, without ever caring for a baby; without ever looking after someone who is ill, old, or lonely; and without comforting or assisting another human being who really needed help. The developmental consequences of such a deprivation of human experience have not as yet been scientifically researched. But the possible social implications are obvious, for – sooner or later, and usually sooner – all of us suffer illness, loneliness, and the need for help, comfort, and companionship. No society can long sustain itself unless its members have learned the sensitivities, motivations, and skills involved in assisting and caring for other human beings. (2001, p. 14)

The Work of Glen Elder Jr.: Central to the work of Elder and his colleagues (Elder 1974, 1986; Elder et al. 1996; Elder 1998; Elder and Giele 2009) is the concept of the *life course*:

> In concept, the life course refers to age-graded life patterns embedded in social structures and cultures that are subject to historical change. These structures vary from social ties with family and friends at the micro level to age-graded hierarchies in work organizations and to the policy dictates of the state. Changes in the life course shapes the content, form, and processes of individual development, and such change may be prompted in part by the maturation or aging of the individual as well as by social forces. (Elder et al. 1996, p. 31)

From this life-course perspective, Elder makes a strong argument for the role of the environment, structured hierarchically in the same sense as that of Bronfenbrenner, and defined in terms of both historical and social structures and events, as a major determinant of human behavior across the entire life span. He does not deny the role of personal choice or agency in this model, but he does make clear that the choices one makes across the life span are largely defined in terms of the options available from which to make these choices.

For Elder, the best way to appreciate the impact of the sociohistorical context is through the study of *cohorts.* Cohorts reflect generations, and the developmental patterns of different generations are, according to Elder, a reflection of the sociohistorical context in which that generation developed. Using the appropriate research designs, it is possible to disentangle cohort effects from two other types of effects, *period effects* and *age-graded effects.* Imagine a series of longitudinal studies, each involving the same age range but begun at 5-year intervals. For example, the children in the first sample might have been born in 1920, reached age 5 in 1925, age 10 in 1930, age 15 in 1935, and, continuing on, age 70 in 1990. The children of the second sample were born in 1925, reached age 5 in 1930, age 10 in 1935, age 15 in 1940, and finally, age 70 in 1995. The children of a third sample would have been born in 1930, reached age 5 in 1935, age 10 in 1940, age 15 in 1945 and they would celebrate their 70th birthdays in the year 2000. There certainly could be more samples and the interval doesn't necessarily need to be 5 years, but for the purpose of defining the importance of cohort in a life-course perspective, three longitudinal samples spaced at 5-year intervals will do just fine.

It is possible to look at the data generated from such a research project from three perspectives. For any give longitudinal sample, I can look at how this group of individuals changed over time, that is, I can document age-graded developmental patterns. How do height and weight change over the life span? What about changes in linguistic or mathematical competence or in skills in negotiating social interactions? If I can show that there is an age-related pattern to these changes, that is, that knowing one's age allows me to predict height or weight or social competence, then I have demonstrated an age-graded pattern.

Because at any given moment in time, I have individuals at different ages (e.g., in 1940, I have a group of 20-year olds, a group of 15-year olds, and a group of 10-year olds), I can also ask about period effects, that is, the impact of some event that has an influence on all of the individuals in each of the successive longitudinal samples irrespective of their ages. Period effects must obviously be of sufficient magnitude as to essentially affect everyone irrespective of developmental level, but such events do occur and, because of their magnitude, typically have lasting effects well beyond the actual event itself. The earthquake and resulting tsunami that struck northern Japan in March, 2011, is a very unfortunate example. The magnitude and scale of the destruction affected everyone, irrespective of age, and the effects of this experience on the lives of those affected will be evident long after the debris is cleared and communities rebuilt.

Cohort effects represent the third perspective and effectively integrate the age-graded and period effects by asking either how different generations react to the same sociohistorical event or how the age-graded developmental pattern changes across generations. In the first case, although it is certainly true that the destruction in northern Japan affected everyone in its path, it is also likely that the nature of this effect differed across age groups. Further, if I find on some measure that those children who were 10 years old in 1930 differed in some systematic way from those who reached the age of 10 years in 1935 who in turn differed from those reaching the age of 10 years in 1940, then I have evidence for a cohort or generational effect. In fact, it was Elder's documentation of significant cohort effects reflecting the impact of both the great economic Depression of the 1930s and World War II that provides the most significant data to support his life-course perspective and, by extension, a nurture position.

Elder was able to compare the developmental paths of three cohorts who experienced the effects of the Great Depression and World War II at different ages. The original work, *Children of the Great Depression* (Elder 1974), followed a single cohort of children from the Oakland, California, area who were born between 1920 and 1922. By virtue of their years of birth, this cohort experienced childhood during a time of economic affluence and an adolescence during a period of what was for many significant economic deprivation. They graduated high school just as the economy was beginning to brighten and just before the start of World War II. Many served in the military and, at war's end, entered their adult years during a period marked by remarkable economic growth, social stability, and, as veterans, ample opportunities for free or low-cost educational advancement.

How did the children of these Oakland families cope with these new life experiences? To what degree, if any, did the Depression leave a legacy? What form might that legacy take? In general, for men, the war and the educational opportunities that followed its end meant that most of the men in the Oakland sample, both economically deprived and nondeprived, acquired good educational experiences and good employment. In this sense, the Depression had little negative impact on the adult work experiences. But at another level,

Elder notes that there were significant differences in the meaning attached to work.

These differences in orientation to work were even evident when these men were in their middle years, ages 45 to 65. Issues of job security and maintaining a reasonable income continued to be much more important to those from families suffering economic deprivation 30 years earlier than to those from relatively nondeprived families. Further, leisure activities meant little to these previously deprived men: Home and work were their lives.

The legacy of the Depression was also significant for women, especially for those from deprived families. Elder found that these women were much more receptive to traditional women's roles as they were defined in the 1940s and 1950s. They married at an earlier age and were more involved in household routines. They were less likely to attend college and, if they did work, were more likely to stop at marriage or the birth of a child. As was true of the men who showed little interest in activities outside of the home and family, for many of these women from deprived families, marriage and family were their lives.

To leave the story of the effects of the Depression at this point might give the impression that there were some significant age-graded effects acting, in both the short and long terms, on those who experienced the 1930s. The issue is, as Elder shows, clearly more complicated. By comparing the long- and short-term impacts of the Depression on two other cohorts, one older and one younger than the people of the Oakland sample, Elder is able to demonstrate the contextual embeddedness of the life course.

Data from two other longitudinal samples were also available to Elder. The Terman sample involved children born between 1900 and 1920 and the Berkeley sample included children born between 1928 and 1929. Because of the differences in the birth dates of these three cohorts, they experienced the same set of events at different points in their lives. The older members of Terman's sample were trying to enter the job market during the depths of the Depression whereas the Berkeley children were in their early childhoods. The members of the Terman sample also served in World War II but, unlike the members of the Oakland sample, their service interrupted already started adult lives. The members of the Berkeley sample were too young to serve in World War II, but many did serve during the Korean War. The members of all three samples experienced the same set of historic events but did so at different points in their lives and by assuming different roles in the family. It is by examining the differences in the experiences of the three cohorts that the power of the sociohistorical context becomes clear.

With respect to the timing of military service, Elder argues that members of the Oakland sample, by entering service as they were making the transition from late adolescence to early adulthood, entered the military at just the "right time." That is, they entered before having established adult careers and at a point in their lives at which they were still deciding on a life direction and at a time when the military offered a dramatic break from the impact of the Depression. Further, for those who were fortunate enough to return from the war, the educational benefits of the G.I. Bill allowed these men the opportunity to act on their newly formed sense of adulthood.

In contrast, the consequences of military service for the older men of the Terman cohort were very different. Because they had already entered their adult years by the start of the war, most had married, many were parents, and all had entered some sort of work or career path. For the men of this cohort, service was more a disruption than a transition point. The return to civilian life was not marked by making use of the opportunities available to start adult roles but rather by efforts to put things back together again. The difference left a legacy. Elder describes the adult lives of the Terman men as punctuated by many difficulties: "They suffered more work instability, earned less income over time, experienced a higher rate of divorce, and were at greater risk of an accelerated decline in physical health by their fifties" (1998, p. 9). This adult life pattern for the Terman men is even more remarkable when you consider the nature of the group. Unlike the Oakland and Berkeley samples, who represented a cross section of the middle- and working-class communities in Oakland. California, at that time, the Terman sample, established by Louis Terman (one of the early proponents of intelligence testing in the United States) was designed to study the development of gifted and very high-intelligence individuals. The people in this sample represented the best of the best.

What about the Berkeley cohort, those who were young children during the Depression? Whereas the children of the Oakland cohort were in their adolescence and were therefore actually able to having a meaningful role in helping their families deal with adversity, the younger children of the Berkeley cohort were still at home, able to do relatively little to help. But by being more present in the home, they were more likely to experience the impact of the family stresses accompanying the loss of income and family disruption. The people of the Berkeley cohort were more likely to be reported as having lower scores on measures of self-confidence during childhood and on measures of assertiveness, social competence, and aspirations during adolescence (Elder and Hareven 1993). Elder believes that these differences reflect the opportunities provided the two cohorts to acquire a sense of human agency. The older children of the Oakland cohort were old enough to be able to do something to help the family cope with its economic losses. These opportunities provided a sense of purpose and meaning and, as such, a sense of self that emphasized human agency. The younger children of the Berkeley cohort, being too young to engage in such purposeful activity, were left to experience the adversity without the opportunity to do much about it.

The members of all three cohorts experienced the same set of historic events but because they did so at different points in their respective life spans, the impact of these events, both in the short and long terms, is significantly different. And although Elder's intent has never been to advocate a nurture position in the classic debate, data such as these do just that:

> Life course theory and research alert us to the real world, a world in which lives are lived and where people work out paths of development as best they can. It tells us how lives are socially organized in biological and historical time, and how the resulting social pattern affects the way we think, feel, and act. All of this has something important to say about our field of inquiry. Human development is embedded in the life course and historical time. Consequently, its proper study challenges us to take all life stages into account through the generations, from infancy to the grandparents of old age. (1998, p. 9)

The Special Case of Socioeconomic Status: Socioeconomic status (SES) occupies a unique place in the nature–nurture classic debate. Both sides offer explanations for why some people find themselves less fortunate than others. Behavior geneticists suggest that, for some, the reason for their economic circumstance is primarily genetic. These individuals simply do not have the basis to successfully acquire the competencies needed to get out of poverty. For those on the nurture side of the debate, the issue is primarily one of social policy and opportunity. In other words, some are less fortunate than others largely because of the particulars of their sociohistorical contexts. The discussion above on behavior genetics elaborated on their explanations of social stratification. What follows offers a nurture perspective on the same issue.

SES is typically defined in terms of years of education, income, and occupational status. But these are social address variables; they really say little about the lived experience of individuals on any rung of the socioeconomic ladder. The real developmental question – the one most relevant to the debate – is how income translate into access, how the nature of one's work experiences influences one's perception of the world, and how one's educational level influences the manner in which information is processed and decisions made. The work of Melvin Kohn (Kohn et al. 1990; Kohn 2006) and the work of Gary Evans (Evans and English 2002; Evans 2004) help to translate these social address variables into such concrete experiences and, in so doing, offer evidence supporting the role of nurture in development.

Kohn's work focuses on the psychological effect of social class. He defines social class in terms of an individual's relationship to ownership and control of the means of production and of one's control over labor. For Kohn then, occupation is the essential defining element of SES, not so much in terms of specific occupations but more so in terms of the distinction between labor and management. The key element is occupational self-direction, the degree to which one is in a position to make decisions directing the means of production and the labor of others. Occupational self-direction, in turn, serves to provide a pattern of experiences that Kohn sees as having broad psychological effects beyond the workplace. In particular, Kohn finds that those in occupational roles allowing a higher degree of self-direction are significantly more likely to value self-direction for their children, to be intellectually flexible, and to be self-directed in their orientation. As such, the impact of occupational role extends beyond the workplace and spans generations, fostering the attitudes, dispositions, and competencies that make social advancement either more or less likely. And in what might seem as a direct challenge to behavior genetic SES arguments, Kohn et al. are quite specific in pointing this out:

> The interrelationship of social class, occupational self-direction, and psychological functioning does not result solely, or even primarily, from selective pressures by which self-directed, intellectually flexible men attain jobs that afford greater opportunity for occupational self-direction and more advantaged class position. Although these processes do occur, the evidence is clear that class position has a powerful effect on opportunities to exercise occupational self-direction, and that occupational self-direction has decided, cross-nationally consistent effects on ideational flexibility, valuation of self-direction, and self-directedness of orientation. Occupational self-direction clearly plays the major role in explaining the effects of social class on these facets of psychological functioning. (1990, pp. 1004–5)

Evans (Evans and English 2002; Evans 2004) takes a broader perspective on SES by considering the physical as well as the psychological correlates of living within different income levels. His research finds that the physical environment of children from low-income households is significantly more likely to involve greater density, higher chronic noise levels, greater exposure to toxins (especially lead), and greater exposure to indoor pollutants such as carbon monoxide and radon; the people in these low-income households are more likely to live in housing with structural problems and ironically, given the condition of the housing available to low-income families, are more likely to have greater difficulty in finding affordable housing.

At the psychological level, Evans reports that children from low-income families are more likely exposed to family violence and disruption. They are more likely to experience inconsistent parenting, which at times may be harsh and punitive, and to have significantly less cognitive stimulation, particularly with respect to language development. These psychological disadvantages may well reflect the social nature of the low-income community because low-income families report fewer social contacts, a smaller support network, and fewer organizational ties.

It is the cumulative impact of these physical and psychological stressors that Evans believes is primarily responsible for differences in the developmental outcomes of children growing up in different income settings. For example, children from low-income households exhibit higher blood pressure and greater evidence of chronic stress hormones, two significant predictors of health issues in later life. Further, these children have greater difficulty in school, are more likely to demonstrate a range of social emotional problems, and are less effective in establishing and maintaining healthy peer relationships, less able to self-regulate their behavior, and typically self-report lower levels of psychological well-being. And, like Kohn, he is clear as to the origins of these SES-related problems. Even when partitioning out essential child and parent characteristics, the wealth of data on SES converges on the conclusion of "poverty adversely impacting children independent of genetic inheritance" (2004, p. 88).

This discussion of SES is by no means meant to be exhaustive; there is obviously a very large literature reporting on the influence of physical and psychological environments on children's development. But it does serve to illustrate what a nurture position offers in attempting to account for the course of development.

Some Tentative Conclusions About Nurture

Environmentalists can be excused if they are feeling a bit defensive these days (Turkheimer 2004). It seems as if every day an article appears in some outlet reporting that someone has "discovered" the gene for something or that scientists have proven what area of the brain lights up when a child does this or that, as if the brain were some sort of pinball machine. Actually, they needn't be quite so concerned because the headlines typically overstate the actual data, but this doesn't mean that some introspection might not be in order. Actually, in some ways the very problem is introspection.

Environmentalists of all persuasions have always tried hard to document the correspondences between definable objective environmental events and patterns of behavior, and they have favored this approach as a strategy to get away from more introspective forms of data, data that are neither easily standardized nor controlled. The behaviorism of the 1950s and 1960s may have been the heyday of such a rigorous, objectivist approach, but the basic value is still evident today. Even when the interpretations of the data are couched in more mentalistic terms such as processing speed or numerical construct, the data themselves are typically still one or more specific behaviors. This strong reliance on observable, quantifiable behavior creates two problems for environmentalists, ones that the other perspectives often use in their critiques of a nurture perspective.

First, as even environmentalists readily acknowledge, there is no agreed-on unit of measurement. This is not necessarily a major stumbling block for any particular research agenda because the unit can be defined with respect to that particular research focus, but it is an issue when trying to generalize across agendas, that is, when trying to get the big picture.

Second, as Bronfenbrenner frequently noted, what are typically defined as measures of the environment are typically measures of social address, such as occupation, years of education, or number of books in the home, than of the processes that take place within those settings. And, although the two do relate to each other, demonstrating a relationship between, for example, family income and parenting strategy still does not explain why having different amounts of income leads to different ways of dealing with children. This is not to suggest that an environmentalist approach is wrong but, as is true of the behavior genetic approach, it can get you only so far, both with respect to accounting for variance and for providing complete explanations of observed patterns. And, as is also true of the behavior genetic approach, here too the approach is necessary but not complete. Just as we can critique a behavior genetic approach for its overemphasis on variance at the expense of measures of central tendency, we can as easily fault the empiricists for just the opposite bias. Somehow the two need to be reconciled.

Some Final Tentative Conclusions

When I was in graduate school, I took courses in research methodology, and one especially stands out in my memory. Professor Raymond Kuhlen was the teacher, and I remember one class in particular when he said in no uncertain terms that all developmental research must have at least three data points, for example, three age groups, to compare. I also remember someone in the class asking Professor Kuhlen why he felt so strongly about this, and he replied that when you have only two, your developmental function will be a straight line, even when it isn't.

I think the researchers involved in the classic debate would have done well to have taken Professor Kuhlen's course because, in effect, they are making a similar error. It isn't an error involving the number of levels of the dependent variable, however, but it is one in their implicit metatheoretical assumptions, namely, that when you assume that variables can be modeled as acting independently of each other, you create research designs and systems of analysis that show that they do, in fact, act independently of each other – and perhaps, as Professor Kuhlen would have been quick to point out, even when they don't. Simply put, this is the fundamental issue with the classic debate. It isn't an issue with the answer, it is an issue with the question.

Behavior geneticists do their research and show clearly that genes have more of an influence on development than environment does. Environmentalists do their research and show clearly that environment has more of an influence on development than genetics does. Both are probably right to one degree or another, and each has helped us better understand some of the factors influencing the human condition. This improved understanding is in no way trivial, but it is limited. It will never provide us the bigger picture, not because, as most journal articles end, by arguing that more research is needed, but rather because a different kind of research is needed. The potential relationship between the two approaches discussed in the next chapter – evolutionary psychology and developmental systems theory – offers just such a possibility.

5

The New Debate

The classic debate is a debate of opposites; the new debate is one of chickens and eggs. On the one hand, the classic debate simply has no resolution because nativists and empiricists cannot even agree on a dependent variable, much less a common methodology. The new debate, on the other hand, is not a dead end; it might just have a resolution. It won't be either the chicken or the egg, however, but it might just be the chick.

This chapter reviews, in turn, the two participants in the new debate, evolutionary psychology and developmental systems theory, and then we see where an integration of the two might just become possible.

Evolutionary Psychology

Most theories of human development focus on proximal variables, ones concurrent with the developmental events themselves or in the very recent past of the individual. This is certainly true of the nativists and empiricists of the classic debate. Evolutionary psychologists see things differently, very differently. Their focus is on the distal antecedents of our behavior and development, causes that are depicted as taking place a very, very, very long time ago.

There is no one theorist who ideas represent evolutionary psychology (EP), as might be the case if we were discussing Piaget or Freud. Rather, there are many people working within this tradition or perspective. Nevertheless, the most consistent and well-developed exposition of EP comes from the series of papers written over a number of years by John Tooby and Leda Cosmides (Cosmides and Tooby n/d; Tooby and Cosmides 1997; Cosmides and Tooby 2001; Tooby et al. 2003; Tooby and Cosmides 2005). I first summarize the model by reviewing their work and then present two variations on the model, one moving EP closer to developmental behavior genetics and the other closer to developmental systems theory (DST).

The core concept of an evolutionary psychological perspective is that it is impossible to understand human development unless that development is viewed from an evolutionary perspective. As such, just as it is argued that the study of all biology must be the study of evolutionary biology, so too must the study of human development be the study of evolutionary human development. Evolutionary psychologists make this claim by arguing that contemporary human activity reflects a brain architecture and functioning that was essentially defined more than 100,000 years ago. In other words, modern humans deal with contemporary problems and tasks using a "stone-age mind." Quite the claim! What is the rationale for the argument and on what support does it rest?

Our earliest hominid ancestors emerged approximately four million years ago. Since that time, several distinct hominid species have evolved and died out. Each successive species is viewed as more evolved than the previous with respect to how similar it was to us. This process of successive ancestors is in fact the process of evolution and reflects the ability of each successive species to adapt more successfully to its surroundings and, in so doing, reproduce more, the bottom line goal of the evolutionary process. Species that do not reproduce successfully, have, by definition, failed to adapt.

What we recognize today as genetically modern humans actually first emerged some time ago – about 100,000 years – and for approximately 90% of our "modern" species history, we lived in small hunter–gatherer bands, numbering probably no more than between 25 and 200. It has been within only the last 10,000 years that this pattern has dramatically changed, mostly seen as being initially due to the domestication of grains that allowed early humans to plant crops and therefore settle. And it has been within only the last 5,000 years that we even have a recorded written history as a species. To evolutionary psychologists, 5,000 or 10,000 or even 100,000 years is too short an evolutionary time frame to argue for significant species evolution. In other words, again, we are what we are largely as a result of the species adaptations made over the several million years before we emerged as genetically modern humans. So what are we?

We are, biologically speaking at least, a set of highly evolved systems for registering and responding to the external environment. The lungs allow for the exchange of air, the stomach for the digestion of food, the kidney for the filtering of waste, and the brain – the organ of interest to evolutionary psychologists – for the processing of information and the regulation of our behavior and physiology. These highly evolved systems came into being through a process of natural selection in which more adaptive strategies (and species) won out over less adaptive ones. These successful adaptations, over a period of millions of years, became coded into our genome and are still seen as regulating our behavior today. One interesting implication of this perspective is that there is no necessary reason to argue that successful adaptations that were made during our hunter–gatherer period are necessarily equally adaptive today. For example, if these adaptations helped early humans live in small groups, then what of our trying to maintain some degree of equilibrium in a world of billions? In fact, evolutionary psychologists argue that many of the problems of modern life can be understood as just such a disconnect.

Evolutionary psychologists view our cognitive adaptations as, in effect, highly specialized organic computer programs that are part of our brain architecture. Further, these adaptations are highly specialized, and, as such, over evolutionary time, we have acquired a large repertoire of such highly specialized task-specific adaptations, typically described as modules:

> To survive and reproduce reliably as a hunter-gatherer required the solution of a large and diverse array of adaptive information-processing problems. These ranged from predator vigilance and prey stalking to plant gathering, mate selection, childbirth, prenatal care, coalition formation, and disease avoidance. Design features that make a program good at choosing nutritious food, for example, are ill suited for finding a fertile mate or recognizing free riders. . . . Domain-specific programs organize our experiences, create our inferences, inject certain recurrent concepts and motivations into our mental life, give us our passions, and provide cross-culturally universal frames of meaning that allow us to understand the actions and intentions of others. They invite us to think certain kinds of thoughts; they make certain ideas, feelings, and reactions seem reasonable, interesting, and memorable. (Tooby and Cosmides 2005, p. 18)

How are these adaptations acquired, how do they eventually become coded into the genome, and how do they regulate behavior? Tooby and Cosmides (2005) see two factors as influencing evolutionary change: chance mutation and natural selection. Of the two, natural selection is seen as the more vital force because it is more likely to increase the survival of the species whereas mutations more often than not are seen as deleterious to the species. Given the fact that we lived as hunters–gatherers for countless generations, the same set of problems was encountered repeatedly across this extended period. Food had to be acquired and determined if edible, decisions had to be made as to whether someone was friend or foe, fertile mates had to be selected, children reared, and predators repelled, to name just a few. Those who demonstrated more effective strategies for dealing with each were more likely to survive and prosper, that is, reproduce. The process did not happen all at once but rather over extended periods and, in the same sense that we would recognize our modern eye as evolving in steps over an extended period of time until all the elements of the design become integrated to form what we now recognize as the modern human eye, so too are these behavior adaptations seen as slowing evolving as a set of "design elements" that eventually become functionally integrated to form an adaptation. Because our ancestors had to confront many different challenges, many different and functionally specific adaptations evolved. The actual number seems to be a continuing matter of debate among evolutionary psychologists, but they are often discussed in terms of hundreds or even more.

The set of challenges early hominids encountered and the context in which these challenges existed are referred to collectively by Tooby and Cosmides as the *Environment of Evolutionary Adaptedness* (EEA). The EEA is not a particular geographic place. Rather it is, in their words, "a statistical composite of the enduring selection pressures or cause-and-effect relationships that pushed alleles underlying an adaptation systematically upward in frequency until they became species-typical or reached a frequency-dependent equilibrium" (2005, p. 22).

It should also be noted that although much of our behavior reflects the evolutionary process of adaptation, not all of our behavior can be classified in this manner. Tooby and Cosmides also distinguish the by-products of adaptations and what they refer to as evolutionary "noise." Reading and writing are examples of such contemporary by-products. Neither is considered an adaptation because there were no ancestral selectionist pressures to develop either skill. Rather, most likely, the adaptation for language has also allowed us to learn to read and write, although not necessarily as well as we learn to talk. Our virtual ease in learning our language of origin as young children is seen as reflecting the adaptive value of such a long-standing skill, but learning to read and write is a much more difficult skill because we have not developed evolutionary adaptations to do either. Instead, each has to be learned though some process of formal instruction, and rarely are we as proficient in either as we are in speech.

Evolutionary noise seems to refer to variability in a species' behavior that does not appear to have much of anything to do with the adaptedness of the species. The variability associated with evolutionary noise would have high hereditability because it is not present in all members of the species and, as such, would be the stuff of behavior genetics. Evolutionary psychologists are interested in only those characteristics that would have low hereditability because such characteristics are species-typical universals and therefore do not vary across members of that species.

Adaptations are not simply cognitive in nature; they are not simply a matter of acquiring specific bits of information. They are equally a matter of knowing when and when not to use this information. In other words, adaptations are as much about motivation and evaluation as they are about content. In fact, evolutionary psychologists see these motivational–emotional–valuation systems as a superordinate component of brain architecture and function because they serve the purpose of eliciting, prioritizing, and coordinating specific adaptations:

> Thus, by evolved design, different content domains should activate different evolved criteria of value, including different trade-offs between alternative criteria. Cases of motivational incommensurability are numerous and easily identified via careful analysis of adaptive problems. Distinct and incommensurable evolved motivational principles exist for food, sexual attraction, mate acquisition, parenting, kinship, incest avoidance, coalitions, disease avoidance, friendship, predators, provocations, snakes, spiders, habitats, safety, competitors, being observed, behavior when sick, certain categories of moral transgression, and scores of other entities, conditions, acts, and relationships. (Tooby and Cosmides 2005, p. 49)

So, what is being proposed by the evolutionary psychologists is a highly evolved and elaborate information processing system made up of a large number of distinct adaptations developed through a process of natural selection housed in the architecture of the brain and coordinated and integrated through a motivational–emotional–valuation superordinate system. Tooby and Cosmides do not make claims as to the literal location in the brain of each of these adaptations and in fact suggest that, rather than each being located in some specific place, each adaptation may be distributed in some fashion in the same sense that our sense of vision is not located in one particular place but rather reflects the coordinated activity of many distinct elements of our neural architecture.

Tooby and Cosmides argue that their model offers a holistic, interactive perspective on the role of nature and nurture in the evolution of adaptations and go to some lengths to distinguish their perspective both from behavior genetic models and from what they refer to as the *Standard Social Science Model* (SSSM), that is, empiricist environmental models. They argue that the relationship is not one of nature *and* nurture (implying the two as distinct) but rather one in which "evolution acts *through* genes but acts on the *relationship* [italics in originals] between genes and the environment, choreographing their interaction to cause evolved design" (2005, p. 35). In other words, at the same time as existing levels of adaptations for any given species act on the environment presumably as effectively as possible, the environment, through the process of natural selection, is acting on the organism by selecting those behaviors that are most likely to give the organism a reproductive advantage:

> Step by step, as natural selection constructs the species gene set (chosen from the available mutations), it selects in tandem which enduring properties of the world will be relevant to development. Thus, a specie's *developmentally relevant environment* [italics in original] – that set of features of the world that a zygote and the subsequently developing organism depend on, interact with, or use as inputs – is just as much the creation of the evolutionary process as the genes are. Hence, natural selection can be said to store information necessary for development both in the environment and the genes. (Tooby and Cosmides 2005, p. 35)

From this perspective, the genes are seen not as the blueprint or instructions of development but rather as the storage site for some information, which, when combined with the information within the developmentally relevant environment, creates the organism's development. In a sense, this conceptualization parallels the process of meiosis in which both sperm and egg each have some of the necessary information but each alone is incomplete to form the organism. In both cases, it is only when the two combine that development emerges.

Not only is the developing organism dependent on a set of genes to determine the formation of structure and function, that is, necessary adaptations, but so too is it equally dependent on the developmentally relevant environment to make the process effective. Neither is seen as having priority over the other, although the cross-generational mechanisms for each differ. For genes, they are passed from generation to generation in the proximate term and change in frequency over the very distal term. For the environment, it is not literally inherited but simply endures across generations and, for some functions, across eons. Organisms exist in a three-dimensional space, with cycles of light and dark, in the presence of other social beings, need to be nurtured when young, and so forth. These things do not change; they are seen as constants and are seen as just as essential to the construction of the organism's adaptations as the makeup of that organism's genotype. Note that this is not a matter of the environment simply enabling the genotype to express itself as might be the case when a plant needs water to grow, but the water doesn't determine whether the plant will be a rose or a petunia. Rather, to Tooby and Cosmides, there is a mutual coordinated, evolved, interdependence between the two such that a change in either element through perhaps mutation or drastic environmental change leads to a disruption of the coordinated developmental process and, as a result, a less viable organism or, in the case of a massive environmental disruption as might be the case during a prolonged war or natural disaster, a less viable species. It is in this sense that evolutionary psychologists sometimes argue that many of society's contemporary ills can be seen as the result of just such environmental disruptions.

To further make this coordinated argument clear, Tooby and Cosmides go to some lengths to differentiate EP from either nativist or empiricist arguments. The primary point of difference with respect to nativists' arguments is that whereas nativists place great emphasis on species diversity, evolutionary psychologists place great emphasis on species-typical universals. And whereas empiricists place emphasis on a generalized, nonmodular model of brain architecture and function, evolutionary psychologists place great emphasis on a highly evolved, multicomponent, highly modularized model.

Humans, and other species for that matter, are very complex organisms. The evolved adaptations that allow our reproductive success require the interdependent functioning of thousands of genes in coordination with a developmentally relevant environment. If humans reproduced by cloning, guaranteeing continuity of reproductive success across generation would not be much of a problem. Barring mutations, each generation would be the same as the previous one. But that isn't the way we reproduce, and so to ensure that each successive generation acquires the necessary adaptations even though their specific genetic makeup is totally unique, evolutionary psychologists argue that natural selection has favored uniformity over diversity of developmental forms simply because uniformity is the more effective strategy to ensure species survival. They do not deny that there are developmental differences; they just argue that such differences are of little, if any, adaptive significance, that they truly are simply noise in the system:

> Thus, humans are free to vary genetically in their superficial nonfunctional traits but are constrained by natural selection to share a largely universal genetic design for their complex, evolved functional architecture. Even relatively simple cognitive programs must contain a large number of interdependent processing steps, limiting the nature of variability that can exist without violating the program's functional integrity. The psychic unity of humankind – that is, a universal and uniform human nature – is necessarily imposed to the extent and along those dimensions that our psychologies are collections of complex adaptations. In short, selection, interacting with sexual recombination, tends to impose at the genetic level near uniformity in the function design of our complex neurocomputational machinery. (Tooby and Cosmides 2005, p. 39)

The only exceptions to this universal pattern that the evolutionary psychologists note are the variability of our blood groups and the fact that we have two alternative sexual designs – male and female. In the case of blood groups, the variability is seen as "playing the odds," that is, making sure that no one infectious disease does us all in at the same time. In the case of sexual form, the differences reflect the specific nature of reproductive-related processes, although even here it is noted that males and females have much more in common genetically than we do differences.

This comparison with nativist models raises an interesting theoretical question, however. Is the diversity associated with evolutionary noise utterly insignificant because it simply is utterly insignificant or is it so only in the time frame of evolutionary processes? Put another way, what, if anything, is the big deal about the mean scores of two groups differing one standard deviation in the greater scheme of things?

Even though Tooby and Cosmides in particular, and their evolutionary colleagues in general, do make the point that EP is different from a behavior genetic view, they do seem to go to even greater lengths to note the distinctions between their perspective and that of the empiricists – so much so that they even feel the need to characterize the empiricist perspective with the label of the *Standard Social Science Model* (SSSM). Defining empiricist approaches as the "standard," evolutionary psychologists in effect are creating a David-and-Goliath–type scenario, and we all know how that encounter ended.

It isn't just that evolutionary psychologists differ with the SSSM; they see it as wrong, radically misconceived, and even as perpetuating a vast tide of human suffering. For the evolutionary psychologists, the SSSM has done all this because of its denial of our basic, evolved nature. Instead, they argue, the SSSM has promoted a view of human psychological architecture that is characterized as consisting of open-ended, generalized, content-independent and equipotential learning and reasoning mechanisms, that is, a blank slate (Pinker 2002). By making the claim that the SSSM perpetuates an image of human psychological architecture as a blank slate on which anything can be written, the evolutionary psychologists argue that we have denied our evolved nature and in so doing have fostered developmental models and social policies and programs that are not only misguided and inconsistent with the data but, because they fail to recognize this evolved nature, have in fact led to human suffering.

Why has the SSSM been so dominant if the evidence supports a more modularized, evolved view of human psychological architecture? Tooby and Cosmides suggest three reasons. First, institutions have vested interests and as such are resistant to change, even when presented with contradictory data. Second, those favoring a SSSM don't really understand an evolutionary psychological perspective, often equating it with behavior genetic models. Third, advocates of the SSSM equate evolutionary arguments with fixed, rigid, behavior patterns of the "me Tarzan, you Jane" variety, a charge that Tooby and Cosmides argue is simply incorrect.

There may actually be a fourth reason, however, for empiricists' rejection of evolutionary theory, and this is the fact that Tooby and Cosmides refer to evolutionary modules as innate *instincts.* For empiricists, the concept of instinct has a long and checkered history and conjures up images of moths throwing themselves at lightbulbs. The evolutionary psychologists say such images are misguided and imply dichotomies (e.g, innate versus learned) that have no place in an evolutionary account of human psychological architecture:

To say that a behavior is learned in no way undermines the claim that the behavior was organized by evolution. Behavior – if it was learned at all – was learned through the agency of evolved mechanisms. If natural selection had built a different set of learning mechanisms into an organism, that organism would learn a different set of behaviors in response to the same environment. It is these evolved mechanisms that organize the relationship between the environmental input and the behavior output and thereby pattern the behavior. For this reason, *learning is not an alternative explanation to the claim that natural selection shaped the behavior,* although many researchers assume that it is. The same goes for culture. Given that cultural ideas are absorbed via learning and inference – which is caused by evolved programs of some kind – a behavior can be, at one and the same time, *cultural, learned, and evolved* [italics in the original]. (Tooby and Cosmides 2005, p. 32)

How do evolutionary psychologists reach such conclusions? Archeologists at least are able to find fossil fragments on which to base their evolutionary hypotheses, but behavior leaves no trace and neither does the soft tissue that makes up the brain. Three strategies are used to triangulate the evolutionary data (Schmitt 2008). One is the logical process of reverse engineering that hypothesizes about what ancient hominids encountered in their evolutionary histories and therefore what adaptations would have needed to evolve to make the species viable. The second is to make testable predictions about what contemporary human behavior would be like if in fact we are still operating with essentially a stone-age mind. And the third is studies of members of modern-day hunter–gatherer societies who, presumably, continue to live their lives much like all of us once are seen as doing.

When engineers are asked to design a solution to a problem, they identify the specifics of the problem, determine available resources, note restrictions on possible solutions, and then design, if possible, a solution that best addresses the problem. Reverse engineering in effect does the same except that instead of coming up with a novel solution, it comes up with what is considered to have been *the* solution, that is, the adaptation. The effort is not simply an 'armchair' exercise because evolutionary psychologists argue that they do in fact know quite a bit about the circumstances in which early hominids lived and knowing the circumstances gives them insight into what adaptations would need to have been made for the species to continue to be viable:

> Ancestral hominids were ground living primates; omnivores, exposed to a wide variety of plant toxins and meat borne bacteria and fungi; they had a sexual division of labor involving differential rates of hunting and gathering. They were mammals with altricial young, long periods of biparental investment in offspring, enduring male–female mateships, and an extended period of physiological, obligatory female investment in pregnancy and lactation.... They lived in small, nomadic, kin-based bands of often 20 to 100; they would rarely (if ever) have seen more than 1,000 people at one time; they had only modest opportunities to store provisions for the future; they engaged in cooperative hunting, defense, and aggressive coalitions; and they made tools and engaged in extensive amounts of cooperative reciprocation. When these parameters are combined with formal models from evolutionary biology and behavioral ecology, a reasonably consistent picture of ancestral life begins to emerge. (Tooby and Cosmides 2005, p. 24)

Reverse engineering, like all forms of retrospective analysis, is not without its critics. Because predicting the past appears easier than predicting the future, critics (Gould and Lewontin 1979; Gould 1991; Rose and Rose 2000) argue that such explanations are best described as "just-so stories," a reference to the children's stories written by Rudyard Kipling. As in Kipling's stories, the explanations are seen as being just too good a fit, more logical (or whimsical, at least in the case of Kipling) than psychological. Nevertheless, evolutionary psychologists argue that knowing the ecological context of early hominids does in fact give us a good picture of the challenges they faced and therefore the most likely strategies they used to successfully meet them. And, they argue, there is very good evidence from anthropological research (Lee and Daly 1999; Marlowe 2005) that similar types of adaptation strategies are commonly found among contemporary hunter–gatherer societies.

Finally, if we do approach contemporary issues with stone-age minds, then data on contemporary humans should show a similar pattern to what would be expected through reverse engineering. And here the reported data can often be uncomfortably similar, especially when they come to research on contemporary male–female differences in sexual behavior with respect to criteria such as both long- and short-term sexual partners, preferences for sexually related literature, and risk taking (Davies and Shackelford 2008). Here too the research is not without controversy (Pinker and Spelke 2005) especially because much of the reported data are characterized as highly sexist but, again, to the evolutionary psychologists, such data simply serve to confirm the role that hominid evolution has played in the development of our human psychological architecture.

In considering these evolutionary arguments, especially as presented by Tooby and Cosmides, there is an interesting differentiation that seems to be evident with respect to the nature–nurture debate. On the one hand, as Tooby et al. (2003) note,, the slow, stepwise process of a species acquiring and integrating adaptations over very long periods of time has very much of the interactive flavor that dictates that the role of nature and the role of nurture are in fact not distinct and partitionable but rather fully holistic. In effect, this acquisition process is really not that distinct from the one described by developmental systems theorists except, of course, for the time dimension. However, once the adaptation has become part of the species genome and is clearly reflected in behavior, then the theory seems to more closely resemble behavior genetic arguments as to the expression of these innate modules in the regulation of the behavior, that is, that nature drives nurture. It is this second "phase" of the model that draws the most comment and criticism (Krebs 2003; Lickliter and Honeycutt 2003). This bifurcation is also evident in the two lines of research that follow from the general model, one looking at an evolutional *developmental* psychology (Bjorklund and Pellegrini 2002; Blasi and Bjorklund 2003) and the other focusing on contemporary adult human behavior, particularly with respect to sex differences (Pinker 2002, 2004; Buss 2009).

The study of sex differences from an evolutionary psychological perspective argues that even though for most things men and women have the same set of innate adaptationist modules, when it comes to issues such as mate selection, sexual activity, parenting, long- and short-term relationships, and work history, that is, places where the behavior of men and women are often seen as differing, the two sexes are innately different because of the cumulative evolutionary impact of selective adaptationist pressures. For the most part, these different pressures are seen as reflecting the fact that women, of necessity, are more invested (by choice or not) in child bearing and child rearing and, given the different fertility patterns of men and women, that women, having "less to give" than men, are more choosy with whom they will share their fertility and are therefore more committed to the welfare of their offspring than are men (Trivers 1972). These differences are seen as universal and innate, evident in all cultures and relatively immutable. The degree of difference between men and women may differ to some degree across cultures, but evolutionary psychologists argue that the direction of the effect, for all aspects of human sexuality, is constant across cultures. Males, for example, may be found to be more promiscuous in some cultures than in others but in no culture is it the case that females are found to be more promiscuous. Further, Davies and Shackelford (2008), in a review of the literature, report on studies that find men more willing to have sex with an opposite-sex stranger, that men report a preference for more sexual partners than women for any given time interval, that men place a higher value on the physical attractiveness of a sexual partner than women, and that the sexual fantasies of men differ both in frequency and type from those of women. In particular, men seem to prefer fantasies involving pornography while women appear to prefer fantasies related to romance novels. Even actions such as domestic violence and rape (Peters et al. 2002; Thornhill and Palmer 2004) are seen as evolutionary expressions of a male's effort to ensure that his female partner does not have sex with other men and therefore be assured that the child his mate is carrying or might carry is indeed genetically his, a major goal in ensuring evolutionary survival. Needless to say, such claims are highly controversial and not without significant criticisms (Dagg 2004), but evolutionary psychologists do take pains to point out that what "is" in no way implies what "ought to be." In fact, they argue that the only realistic way to deal with and ultimately eliminate social ills such as domestic violence and rape is to gain a full and scientifically valid understanding of the origins of such behavior:

> Punishments that increase the costs of socially undesirable behaviors such as rape could be used to reduce the likelihood of socially franchised men raping women both in and outside of war. Evolution is morally neutral and therefore cannot be a basis for deciding the moral worth of punishing any behavior. Evolution, however, strongly suggests the idea that punishing a behavior that people consider undesirable will reduce the behavior. Whether to use punishment and how severe it should be has to be decided by voters. However, focusing policy on punishment without policy adjustments in socioeconomic inequality, job opportunities, educational support, and health care probably would be largely futile. Unless a brighter future is envisioned, humans are not expected to forgo risks, delay reproductive effort, and increase allocations to somatic effort. (Thornhill and Palmer 2004, p. 271)

Even at a broader cultural level, evolutionary psychologists such as Buss (2001) argue that our image of cultures as being uniquely individual in character, infinitely variable, and developing in ways that are particular to its context are in fact incorrect. Rather, he argues, cultures, in nontrivial ways, are in fact very much like each other, and the reason for this surprising uniformity is our shared evolutionary history. All cultures have status hierarchies; those at the top of the hierarchy enjoy more resources than those at the bottom, all have an incest taboo, interpret facial expressions in the same way, have a division of labor by sex, favor kin over others, and so forth. These traits are seen as reflecting the innate modules that all humans carry in the genome. Their expression serves to define the constraints on the forms of cultural expression. Buss argues that even the very fact that anthropologists and cross-cultural psychologists are still much more likely to focus on cross-cultural differences than on cross-cultural similarities may itself be an evolutionary reflection of modules that are designed to detect differences, presumably a more useful adaptation, than similarities. Again, in both the discussions on sex differences and cultural universals, the emphasis is on innate modules that, although once acquired through an interactive process, now seem to be a sole reflection of their expression through the genotype. The role of the environment is now simply to provide a setting in which the innate module is expressed. The environment is given little, if any, role in defining what the module expresses. The acquisition and the expression of modules are portrayed as two apparently distinct processes.

This shifting of EP principles, as expressed by Tooby and Cosmides, toward a more nature-oriented behavior genetic perspective is well illustrated in Pinker's writing (2002, 2004, 2005, 2006). While continuing to emphasize the universality of human behavior and while denigrating – in contrast to the behavior geneticists – the significance of race and ethnicity as significant factors in human development, he nevertheless argues that it is both possible and useful to partition nature and nurture, and that, of the two, nature (i.e., innate adaptationist modules) plays the dominant role in our development. Also, in contrast to the behavior geneticists, Pinker does not seem to argue for a one-to-one type correspondence between specific genes and behavior but rather, in keeping with evolutionary perspectives, that genes influence behavior in a more collective, nonlinear fashion. Indeed, the genetic regulation of brain architecture and functioning envisioned by evolutionary psychologists that gives rise to innate adaptations are portrayed in just this nonlinear fashion. But, at the same time, consistent with a behavior genetic perspective, he also argues for a reductionist perspective as the only way to ultimately make sense of the interplay of nature and nurture and strongly criticizes interactionist, developmental systems approaches as, by refusing to acknowledge the value of disentangling nature and nurture, standing in the way of a full understanding of the determinants of human behavior and development.

Both Buss's and Pinker's works emphasize the expression of evolutionary adaptations in adults and, whether by intention or not, these efforts place little emphasis on the interactive nature of the process as described by Tooby and Cosmides. Rather, the presentations give the impression that these adaptations are innate behavioral forms that actuate themselves in appropriate circumstances. But circumstance has little to do with the particular expression of the adaptation; it is simply the trigger for it.

In contrast to those such as Buss and Pinker whose works support a closer link between EP and developmental behavior genetics, the work of Bjorklund and his colleagues (Bjorklund and Harnishfeger 1990a, 1990b; Bjorklund and Pellegrini 2000; Geary and Bjorklund 2000; Bjorklund and Pellegrini 2002; Bjorklund 2003; Blasi and Bjorklund 2003; Bjorklund 2006; Grotuss et al. 2007; Causey et al. 2008) appears to take EP in the opposite direction, moving it closer to DST. Bjorklund does this by focusing more on childhood rather than on adulthood and by arguing that the contemporary expression of evolutionary modules in both children and adults reflects the same interactive process as was involved in their original formation. The difference, of course, is that whereas the original formation occurred over a very long period of time and is a matter of phylogeny, the module's expression takes place over a much shorter time and is a matter of ontogeny.

Bjorklund and his colleagues argue that rather than adaptations simply popping up whole in adulthood, there is a developmental dimension to their expression. As such, they make a distinction between those adaptations that are most relevant to the adult years but need time during humans prolonged period of immaturity to become fully developed (*deferred adaptations*) and those adaptations, such as infant imitation of adults' facial expressions, that are most relevant to the juvenile period because they serve an immediate survival value (*ontogenetic adaptations*). Further, although clearly recognizing the presence and essential importance of innate adaptations, they also argue that evolution has also made possible a significant degree of developmental plasticity so that organisms at all developmental levels can respond effectively to encounters in a species-atypical environment. This emphasis on plasticity suggests that humans may be able to cope much more effectively to the evolutionary "atypicalness" of contemporary society that traditional evolutionary psychological models would predict.

In contrast to the "gene's eye view of psychological functioning" typical of EP, an evolutionary developmental psychology sees psychological functioning as a result of an epigenetic process at all levels and at all stages of development. Rather than the "instructions" for structure and function being coded in genes, the regulation of development reflects this epigenetic interaction that is constrained equally by both environmental as well as genetic factors. As such, uniformity in development – why, for example, we almost always have the right number of fingers and toes – is as much a reflection of the constancy of the prenatal environment for humans as it is the constancy of one's genotype. In other words, it isn't that the genotype needs an environment in the sense that a plant needs water, that is, to allow a presumably precoded set of instructions to operate. Rather, the genotype needs the environment to provide its information to complete the set of instructions for fingers and toes and higher-order cognitive competencies and everything else for that matter. Human adaptive evolution then is better appreciated as the evolution of epigenetic processes rather than simply the evolution of gene frequencies. So, in answer to the question of "what infants are born with," Bjorklund and Pellegrini argue that

> The answer is epigenetic programs that have evolved over eons and are responsive to the general types of environments that our ancient ancestors experienced. It is these programs, in constant transaction with the environment (broadly defined) that produce the patterns of development and eventual adult behavior that defines us as a species. (2002, p. 336)

And it is to these epigenetic, developmental systems models, especially in terms of the pioneering work of Gilbert Gottlieb and his colleagues (Gottlieb 1997, 2003) and, more recently, the psychophysiological work of Michael Meaney and his colleagues (Szyf et al. 2008; Meaney 2010), that we now turn. Once we review these systems models, we'll return to the problem of the chicken and the egg.

Developmental Systems Theory

We all start life as a very small ball of undifferentiated cells, each cell containing the same genetic component. And yet, very soon, it becomes apparent that these identical cells somehow begin to differentiate, begin to form distinct organs and to function in distinct ways. This pattern continues across the life span. Even though each of our cells contains the same genotype, they function in ways specific to the particular organ and organ system that they belong to. How does this happen? How is it that each of us starts as an undifferentiated, unipotential, bunch of cells and ends up as a complex, integrated organism?

For development systems theorists, the answer does not, and in fact cannot, lie in the genome itself because each cell is at first identical. Something else needs to be added to the explanation. Something has to be added that explains how this undifferentiated bunch of cells ends up very differentiated or, as Watters (2006) asks, how do the instructions get their instructions? Additionally, and to complicate matters even further, we don't seem to actually have enough genes to put ourselves together. Recent estimates of the number of coding genes in our genome puts the number at approximately 25,000. To put this in perspective, this places us somewhere between a chicken that has about 17,000 genes and a grape that apparently requires about 30,000 genes to make itself into a good wine (Pertea and Salzberg 2010). And the 25,000 are not even all unique to us; estimates are that we might share, with apologies to each, as much as 35% of our genes with a banana and up to 80% with a platypus! Again, for systems theorists, something else must be happening than just direct gene expression for us to become human.

Developmental systems theory (DST) tries to identify the instructions for the instructions. Like the evolutionary psychological accounts of Tooby and Cosmides (2005), DST is a holistic perspective, arguing that there can be no distinction between nature and nurture because the way development occurs requires an interdependence of the two. They are simply not separable because neither exists in isolation from the other. But, in contrast, DST does not focus on making specific predictions about our behavior but rather on an understanding of the interdependent, multilevel, bidirectional processes that occur between genotype and phenotype that ultimately are reflected in our behavior over time. In other words, for DST, the issue isn't the genotype per se because all cells have the same genotype; rather it is the regulation of the expression of those genotypes, and why that expression differs across cells, organs, organ systems, and individuals.

DST argues that the process of translating genotypes into phenotypes is so complex and so context related that there is virtually no one-to-one correspondence between a particular genotype and its phenotype. As such, DST describes developmental outcomes as probabilistic. The only exception they recognize is when the inheritance of a particular gene or mutation of one or more genes during development creates such a level of pathology that the organism cannot survive. But DST is quick to point out that such circumstances should not be used as a model or illustration of typical developmental processes because, unlike states of pathology, typical developmental processes reflect the interplay of many genetic and epigenetic factors. In fact, it is the equal emphasis placed on epigenetic factors that has led some (Pray 2004; Wong et al. 2005; Goldberg et al. 2007; Jacobson 2009) to use the term *epigenesis* to describe this interdependent, multilevel, bidirectional process. None of this is meant to downplay the role of the genome in regulating the process of development. Rather, the intent of an epigenetic perspective is to recognize the other factors that also influence the process, that is, to create a degree of parity between genetic and epigenetic variables (Oyama 1985; Griffiths and Knight 1998; Greenspan 2001):

Our DNA – specifically the 25,000 genes identified by the Human Genome Project – is now widely regarded as the instruction book for the human body. But genes themselves need instructions for what to do, and where and when to do it. A human liver cell contains the same DNA as a brain cell, yet somehow it knows to code only those proteins needed for the functioning of the liver. Those instructions are found not in the letters of the DNA itself but on it, in an array of chemical markers and switches, known collectively as the epigenome, that lie along the length of the double helix. These epigenetic switches and markers in turn help switch on and off the expression of particular genes. (Watters 2006, p. 33)

The eminent geneticist Conrad Waddington is credited with first using the term epigenetics with respect to development in a paper published in 1942 (Goldberg et al. 2007). Waddington talked about development in terms of an *epigenetic landscape* and portrayed this image as a ball (representing the cell) rolling down a hill that had a series of bumps and turns and gullies. The deeper the gully, the more inevitable the phenotypic outcome for the cell because the depth prevented the ball from moving in any other direction; in Waddington's terms, the outcome was highly *canalized*. Flatter gradients meant that more variability is possible in the developmental outcome, that is, epigenetic factors would play a more prominent role in determining that aspect of the phenotype. In this sense, developmental plasticity (represented by a flat epigenetic landscape) and canalization (represented by a deeply etched epigenetic landscape) were seen as two ends of the same continuum (Jablonka and Lamb 2002).

Although no one single theorist has defined DST, the work of Gilbert Gottlieb (1983, 1991, 1992, 1995, 1997, 2003, 2007) probably comes closest because it has been his work that has done more to define the perspective than that of any other scholar. Gottlieb can perhaps best be described as a biopsychologist, one interested in the bidirectional interplay of an organism's biology, cumulative developmental history, and current status and context. Fully understanding development ultimately requires an understanding of the interdependence of all three, a perspective he credited to the early work of Kuo (1924, 1929, 1976).

Much of Gottlieb's early work focused on offering a counterargument to the perspective favoring instinctual behavior, which Lorenz (1965) and others were actively promoting. Lorenz made clear distinctions between those elements of behavior that were learned and therefore required prior experience for their expression from those that were instinctual and therefore did not (a position still advocated by some evolutionary psychologists). Gottlieb did not agree and developed some remarkable experiments to counter Lorenz's contentions.

Ducks and chickens respond to the call of their mothers as soon as they hatch. Because there was no time for this skill to be learned, Lorenz claimed that such behavior was instinctual, that it was somehow a direct expression of the animal's genotype. Such theoretical partitioning of antecedents like instinct and learning reflected Lorenz's reductionist perspective. Gottlieb did not agree either with his conclusion or the perspective.

In a series of ingenious experiments, Gottlieb was able to demonstrate that experience does play an essential role in ducklings being able to recognize and to prefer their own species call but that the essential experiences occurred prenatally rather than postnatally. Gottlieb found that when he surgically inhibited the duckling's ability to make sounds while still in the egg, when the ducklings did hatch they were less likely to show a clear preference for the maternal calls of their own species. In other words, the prehatched ducklings' ability to exercise their vocal and auditory capabilities is an essential element in the ducklings' ability to then use their mother's call as a way to orient to and approach her, an essential element of the ducklings' survival.

Gottlieb's findings highlight two components of a DST approach. The first is that no distinction can be made between what, for example, Lorenz would refer to as instinctual behavior and learned behavior. To Gottlieb and other system theorists, such distinctions cannot be made because elements of a system acquire meaning or purpose only in relationship to each other. And, in the case of the ducklings, the necessary experience is not easily classified as either learned or innate but is simply a necessary but not sufficient element in the ducklings' eventual ability to follow their mother. The second point to note is that the prenatal auditory stimulation illustrates the fact that "nonobvious" influences can be just as crucial to the development of some competences as what would appear to be more obvious ones (Johnston 1987). In this case, the ducklings must be able to hear the mother's call in order to orient – the obvious component – but apparently must also experience other epigenetic events that helps structure the phenotype to allow them to, once hatched, make the necessary association between the call and the mother:

> The finding of nonobvious experiential bases of unlearned behavior forces us to think in a new way about the role of experience in the development of behavior that is thought of as instinctual. As discussed in chapters 3 and 4, there is not only one role of experience (conditioning or traditional forms of learning) during development, there are at least three others: induction, facilitation and maintenance. The interesting things about these three modes are that (a) they do not fit the definition of traditional (i.e., associationistic) learning, (b) they entail normally occurring specific patterns of stimulation to achieve the species-specific behavioral outcomes, and (c) their role is not necessarily developmentally obvious (i.e., they are not related to the outcome in a straightforward way). (Gottlieb 1997, p. 76)

The conclusion Gottlieb drew from studies such as the one mentioned above was that development is best appreciated as a process characterized by increases in complexity and novelty over time that reflects the cumulative, bidirectional interactions across and within all levels of the organism and the organism's environment from the level of the genotype to the level of behavior. Further, this process results in the "sequential emergence of new structural and functional properties and competencies at all levels of analysis" (1997, p. 126). In other words, for Gottlieb and other systems theorists, the fact that organisms begin as a cluster of equipotential cells, the fact that there is rarely evidence of a one-to-one correspondence between genome and behavior, and the fact that there is clear evidence for varied developmental pathways for both typical and atypical development demand that the explanation of the developmental process be holistic and systemic rather than a reductionist partitioning of nature and nurture. Such *equifinality* of developmental outcome, according to Gottlieb (2007), is reflected in the difficulty of replicating studies, particularly those dealing with psychopathology, which attempt to demonstrate a direct link between a particular genetic marker and a particular pathology such as schizophrenia. For Gottlieb, the reason for the difficulty in replication is that such links are not inevitable but in turn depend on the present or absence of other factors at any one of a number of levels of analysis from gene to anatomical structure and function, to physiological regulatory mechanism, to behavior, and finally to environment context. Just such a pattern of results is evident in the work of Caspi and his colleagues (Caspi et al. 2002; Moffitt et al. 2002; Caspi et al. 2003) in their studies of childhood maltreatment and depression in male young adults.

In their studies of the effects of childhood maltreatment, Caspi et al. (2002) found that there was neither a main effect for maltreatment on adult criminal behavior nor a main effect for any genetic marker on adult criminal behavior. Rather, childhood maltreatment predicted adult criminal behavior only in children who have a genetic deficiency in the MAOA gene. The MAOA gene influences the metabolism of the neurotransmitters norepinephrine and dopamine, both of which influence an individual's ability to deal effectively with stress. In particular, maltreated children with low MAOA activity were most likely to later show adult criminal activity whereas those maltreated children with normal MAOA activity showed no such association. In another study, Caspi et al. (2003) reported a similar interaction involving depressive symptoms in adults and genes regulating serotonin uptake in the brain. Serotonin also influences an individual's ability to deal effective with stressful life events. Low levels of serotonin were found to predict depression in adults who had experienced stressful life events but not in adults who lives were less stressful. Similarly, holding stress levels constant, only individuals with the allele for inefficient serotonin transcription demonstrated depressive symptoms. The conclusions of Caspi et al. make very clear the value they see in a systems approach:

If replicated, our GxE findings will have implication for improving research in psychiatric genetics. Incomplete gene penetrance, a major source of error in linkage pedigrees, can be explained if a gene's effects are expressed only among family members exposed to environmental risk. If risk exposure differs between samples, candidate genes may fail replication. If risk exposure differs among participants within a sample, genes may account for little variation in phenotype. We speculate that some multifactorial disorders, instead of resulting from variations in many genes of small effect, may result from variation in fewer genes whose effects are conditional on exposure to environmental risk. (Caspi et al. 2003, p. 389)

Belsky and Pluess's (2009) research on plasticity in early human development further helps to illustrate such systemic interdependencies. Belsky and Pluess note that resilience to stress, generally considered an advantage in development, should really be thought of in terms of openness to experience. Although it is true that less resilient children are more vulnerable to developmental risk, it is also the case, they report, that these same children may actually have a developmental advantage in less stressful environments, that is, the same greater plasticity that makes them more vulnerable to risk also makes them more able to benefit from positive developmental experiences. In his words, these are children who are *differentially susceptible to environmental influences*. Belsky and Pluess note, for example, that not only did Caspi's individuals with low serotonin transporters report greater depressive symptoms when exposed to risk but these same individuals reported *fewer* symptoms than the individuals who were the controls when reared in a highly supportive early environment. Belsky and Pluess's research can serve to highlight Gottlieb's bidirectional developmental model by suggesting that, for example, prenatal stress should not be viewed as inevitably a negative antecedent but rather as something that potentially increases the plasticity or opening of the individual to all postnatal developmental events. When these postnatal events are negative, then these children will show greater deficits, but when the experiences are positive, these same children may demonstrate just the opposite pattern.

So how in fact does this multilevel, bidirectional, probabilistic developmental system work? What are the actual mechanisms that regulate the developmental process? What are the instructions for the instructions and where do they come from? Gottlieb (2007) and other systems theorists are quick to acknowledge that much yet needs to be done to define these developmental pathways, that in fact the surface has barely been scratched. But recent epigenetic work is beginning to make this admittedly abstract metatheory more concrete.

As described by Jablonka and Lamb (2002), epigenetics is concerned with those biological systems that regulate gene expression in the phenotype. These biological systems are viewed as partially self-organizing mechanisms that are particularly sensitive to the physical and chemical properties of the internal and external environments. Epigenetic research builds on the fact that only a small percentage of our DNA actually serves to code for specific proteins. The larger percentage of our DNA serves to regulate those genes that do code for structure. Further, these regulatory genes are particularly sensitive to influences at all levels of the biological and environmental context. Epigenetic research therefore focuses on identifying these regulatory mechanisms. One important implication of an epigenetic perspective is that any given genotype may eventually produce a wide variety of phenotypes, depending on the context in which these genes are expressed (Gottesman and Hanson 2005). The variety is far from infinite; there clearly are species-specific genome-related developmental limits, but an epigenetic perspective is making it increasingly clear that these limits are much wider than more traditional reductionist approaches claim. And perhaps no one's work make this clearer than the research program of Michael Meaney and his colleagues (Lupien et al. 2000; Meaney 2001, 2004; Meaney and Szyf 2005; Szyf et al. 2008; Meaney 2010), in particular their work on the effects of maternal care on gene expression and phenotypic development in mammals.

A female rat gives birth to litters of about ten to fifteen pups and is the sole source of nutrition before weaning. During this interval, Meaney reports that the life of the female consists solely of a cycle of nursing and then replacing nutrients so that nursing can again occur. Nursing begins when the female gathers her pups and begins to groom them through licking. The grooming arouses the pups such that they begin to actively nurse, which in turn ensures a milk letdown and the pups are fed. This process of licking and sucking continues even when the milk has been exhausted. Meaney notes that about a third of the licking is directed at the anogenital region of the pup. This focus on the anogenital region serves the important function of ensuring that the pups will urinate. And unlike the parents of most human infants, who no doubt secretly hope that their child will never urinate again until fully toilet trained, the female rat is very pleased to see her pups urinate because it is the high sodium levels in the pups' urine that is her primary source of nutrition during this early postnatal period.

But not all mothers lick and groom (LG) equally – high-LG mothers do so about three times as frequently as low-LG mothers. And the differences are significant in terms of the development of the offspring. Pups of high-LG mothers are, as adults, less fearful in stress situations than the pups of low-LG mothers. Of particular importance is the fact that this link between postnatal maternal behavior and adult pup behavior is mediated by the functioning of the hypothalamic–pituitary–adrenal (HPA) axis. The HPA axis is a major component of the neuroendocrine system and serves to control, among other things, an organism's response to stress. Further, when offspring of low-LG mothers are placed with high-LG mothers, their HPA axis when adults is similar to that of high-LG offspring, that is, they are also less reactive to stressful events. But how does this happen? What is the mechanism through which maternal grooming during the neonatal period could possibly affect gene expression in the adult?

Maternal licking and grooming activates a cascade of neuroendocrine activity that serves to alter the responsiveness of DNA to epigenetic factors. The DNA itself is not changed; its regulation is changed. In this case, the change serves to inhibit the expression of gene activity that otherwise increases the pups' response to stress. The process is referred to as DNA *methylation.* DNA methylation regulates the regulation of gene expression in that it influences the receptivity of the gene to other epigenetic variables. In particular, the mother's behavior alters epigenetic markers on DNA that in turn regulate the transcription of, in this case, the glucocorticoid receptor, which in turn regulates the HPA response to stress. Meaney argues that this description of DNA methylation regulating stress response in rats is not unique to this one stress response system, but rather that it reflects the more general case of epigenetic factors regulating gene expression throughout the organism. And the methylation effects seem to last across generations, not by changing the genome but through epigenetic channels. In other words, the effect of placing a low-LG pup with a high-LG mother continues across generations because the female pups, when they parent, demonstrate high-LG behavior, which in turn serves to regulate neuroendocrine activity of now the third generation.

Meaney sees this cross-generational pattern as highly adaptive because the behavior of parents, intentionally or otherwise, serves to prepare their offspring for the same environment as that of the parent. In the case of rats, high-stress environments produce low-LG mothers who because of their lesser "involvement" in their offspring's early development, produce rats who are more reactive to stressful events, the very events that have a high probably of occurring in those rats' environment. The pattern would be the opposite for rats reared in low-stress environments. Their mothers would be high LG, leading to less stress reactivity in the offspring. Meaney suggests that similar epigenetic-context-related neurendocrine mechanisms are probably equally present for humans growing up in high-stress versus low-stress environments:

> In this environment, the shier and more timid males were more successful in avoiding the pitfalls associated with such "criminogenic" environments. Under such conditions a parental rearing style that favored the development of increased stress reactivity to threat would be adaptive. Thus it is understandable that parents occupying a highly demanding environment might transmit to their young an enhanced less of stress reactivity in "anticipation" of a high level of environmental adversity. Such a pessimistic developmental profile would be characterized by an increased corticotropin-releasing factor gene expression, and by patterns of gene expression that dampen the capacity of inhibitory systems, such as the hippocampal glucocorticoid receptor system.... The obvious conclusion is that there is no single ideal form of parenting: various levels of environmental demand require different traits in the offspring. This is a simple, even obvious message, with significant social implications. (Meaney 2004, p. 13)

The significance of the works of Gottlieb and Meaney and other developmental system theorists should make clear why DST argues that efforts to differentiate nature and nature are meaningless or, as Meaney (2010) expresses it, "biologically fallacious." The issue is not that genetic and epigenetic actions and interactions are so complex as to make it difficult or even impossible to differentiate nature and nurture; rather it is that there is no differentiation because neither exists independent of the other. Elements of a system exist only in relationship to other elements of that system. Any attempt to study one element independently of all others is therefore impossible because that element exists only in relationship to all other elements. This statement is particularly true when gene expression is concerned (Burian 2005). What a gene does, that is, what protein is formed, depends on when it is expressed and how it is spliced together with other DNA strands. The same strands in different combinations in fact code for different RNA transcriptions, and all of this activity in turn is dependent on the presence and activity of other genes, on what previous gene-related activity has occurred in the organism, and on the variety of biological activities occurring at all levels of the individual and the individual's context. In other words, from a system's perspective, there simply are no one-to-one correspondences between gene and behavior or, equally, between environment and behavior. It just depends.

The Chick

Evolutionary psychology (EP) deals with units of time measured in eons; DST in units of years. EP ultimately is concerned with phylogeny, the evolution of species; DST is concerned with ontogeny, the development of the individual. Where could there possibly be a fit? Well, truth be told, there couldn't be until fairly recently, specifically until the *modern synthesis* in biology began to be questioned.

The modern synthesis in biology has defined the study of evolution and genetics for most of the last century and represents an integration of Darwin's views on natural selection, Mendel's laws of inheritance, and molecular genetics. In essence, the modern synthesis saw natural selection as the primary mechanism (along with genetic mutation) for genetic evolution. The evolution of a species became the evolution of its genome, and the genome was seen as the hardwired code for the development of the organism; hence the search for "genes for." The modern synthesis depicted phylogeny and ontogeny as distinct, and, as such, molecular geneticists and developmental biologists (née embryologists) weren't seen as having much to say to each other, at least as far as the molecular geneticists were concerned. The reason was simple: Phylogeny, according to the modern synthesis, influences ontogeny through the evolutionary process but ontogeny does not influence phylogeny. In other words, ontogeny was no more than the transcription of a predefined genetic code and so, from an evolutionary perspective, the study of ontogeny held less value to understanding a species than did the study of its phylogeny. The idea seemed solid enough until we developed better ways of actually counting genes, and the more accurate the counts became, the fewer the number of genes there were to be found, clearly fewer than what would be needed to support a "genes for" view of development. Even Gottlieb, in his 1997 book, estimated that there were approximately 70,000 unique genes, yet only a few years later when the Human Genome Project reported its findings, the number now stood at approximately 25,000. Something was clearly amiss. There had to be more to the process than just genes coding for something. There simply weren't enough genes to code for everything, especially when simpler organisms and even the humble grape had larger genomes than we do.

The solution has been to look at epigenetic factors, exactly the type of work that Meaney does. But epigenetic factors suggest the phylogeny–ontogeny link may be bidirectional, and as such the modern synthesis may be in need of revision. The epigenetic paradigm has certainly had a significant effect on biology, especially in terms of an increasing rapprochement between molecular geneticists and developmental biologists, so much so that a new subspecialty linking the two is now referred to as *evo–devo* (Arthur 2002; Hall 2003). And a similar rapprochement might just be at hand with respect to evolutionary psychologists and developmental systems theorists, and this gets us back to the chick:

> Epigenetic inheritance also means that the distinction between developmental (proximate) causes and evolutionary (ultimate) causes is not as clearcut as we have been accustomed to believe, because developmentally acquired new information can be transmitted. Proximate causes are sometimes also direct evolutionary causes. The closely related assumption that instructive processes (processes that lead to induction of the functional organization of a system) are the subject matter of development while selective processes (those that "choose" among variant systems) are sufficient to explain evolution, also need to be modified. If development impinges on heredity and evolution, then there are some instructive processes in evolution too. It follows from that that the distinction between replicator and vehicle, or even replicator and interactor, is in many cases inappropriate. (Jablonka and Lamb 2002, p. 94)

A focus on epigenetic mechanisms gets us to the middle, to a place that focuses on how phylogeny influences ontogeny and, at the same time, ontogeny influences phylogeny. That place is, of course, early development, the place between the species and the adult, that is, the chick. It isn't a necessarily easy place to get to, especially when talking about the relationship between EP and DST because there has been a tendency for each to paint the other in extreme terms, terms that serve to push each to the border and away from the center. But this too may be changing.

A good illustration of the problem of getting to the middle is the article by Lickliter and Honeycutt (2003) and the subsequent rebuttal by Tooby et al. (2003). In their article, Lickliter and Honeycutt argue that EP presents an inaccurate picture of development. First, they argue that, as is equally true of sociobiological accounts of development, EP presents a gene-centered account of both phylogeny and ontogeny, that is, the instructions for all aspects of the organism, including morphology, physiology, and behavior, are present in the gene. Second, they fault EP for presenting an image of ontogeny that is little more than the predetermined unfolding, as a result of the phylogenetic history of the species, of a set of instructions coded within the genome. Such a predeterministic perspective, according to Lickliter and Honeycutt, "assumes that development is internally determined, set on a course at conception, and specified by genetic programs designed and selected over evolutionary time" (2003, p. 821). And the role of the environment, presumably, is little more than that of a trigger or activator. Third, Lickliter and Honeycutt find fault with EP's argument that natural selection has crafted a set of innate modules, ordering into the hundreds or even thousands, into the genome that serves to solve a specific adaptive problem encountered in our evolutionary past. They argue that such modules are seen as having a predetermined structure of innate knowledge and an innate set of procedures for applying this knowledge in the appropriate setting, that is, see hungry lion, run fast in opposite direction. And finally, they argue that EP offers an outdated conception of the gene, in particular that EP talks "about them as if they have persisted in immutable form, impenetrable to outside influence for millions of years" (2003, p. 823). Lickliter and Honeycutt then go on to offer an alternative developmental systems view as was discussed in the previous section. In their conclusion, they summarily fault EP for offering a view of behavior and development no longer tenable given our current level of knowledge in the biological and psychological sciences. In particular, they argue that EP offers, contrary to current evidence, "an unnecessarily reductionist view of the emergence and maintenance of phenotypic traits by treating genes as causal agents with closed programs" (2003, p. 830). Needless to say, Tooby et al. saw the situation differently.

In their rebuttal, the three authors argue that either the critiques made by Lickliter and Honeycutt are simply not true or that they are trivial. So, for example, Tooby et al. argue that the claim that EP is insensitive to epigenetic influences is "to put it mildly, not a new insight for evolutionary psychology but instead the starting point for their research" (2003, p. 858). Further, in comparison with DST's "bland general statements (true of all organisms at all times)," EP, by empirically testing its hypotheses, is able to offer – and confirm – a number of specific predictions about individual behavior in particular settings, something DST does not and, according to Tooby et al., cannot provide:

> In comparison, developmental systems theory makes no specific predictions of any sort and thus is useless as a scientific theory. Its uncontroversial assertion that everything potentially interacts with everything else – and that developmental outcomes are dependent on a convergent nexus of joint determination that varies depending on the case – is compatible, after the fact, with any imaginable research finding. It is, indeed, a statement of an absence of (and an aversion to) principled knowledge, disguised as a theory. (Tooby et al. 2003, p. 860)

Tooby et al. also challenge Lickliter and Honeycutt's claim that EP sees no role for the environment in development other than as the trigger for the predetermined unfolding and expression of genetically coded modules. Quite to the contrary, they argue that EP sees the environment as occupying as central a role in defining development as it does the genome. In particular, what they refer to as the *developmentally relevant environment* is viewed as a second system of inheritance that, through its stability and along with the genome, regulates behavior and development across generations. Environmental change is as disruptive of inheritance across generations as is genetic change:

> The species-typical features of the genome interact with the features of the evolutionary long-enduring, special-typical environments to produce the species-typical design observable in all of us. Failures of reliable development are attributable to genetic mutation, to environmental mutation (change) or both.... The fact that we have all been here before is a very good thing, because it is only through this regularity of interaction, from generation to generation, that allows organisms to climb toward a tolerable level of functional organization and inclines them away from depending on the aspects of the environment that are unpredictable variable and hence disordering. As evolutionary psychologists, we believe in design reincarnation based on two inheritances, not genetic preformationism based on one. (2003, pp. 863–4)

If we take the critique and rejoinder at face value, there would appear to be little that EP and DST have in common. But if we look carefully instead at what the advocates for each perspective argue, then there clearly are points of convergence. In particular, both perspectives seem to recognize that there is an interplay between ontogeny and phylogeny that most likely reflects the actions of epigenetic mechanisms across generations; that natural selection acts on organisms (i.e., phenotypes) as they attempt to adapt to environmental conditions; that attempts to differentiate nature from nurture have no theoretical basis; and that the distinction between what might be seen as instinctual or innate and what might be seen as learned or acquired is probably more apparent than real.

Both perspectives recognize the crucial role that adaptation plays in the evolutionary process. Rather than evolution being seen primarily as the result of chance, rare mutations in the genome, the efforts of the organism itself to adapt seems to have an influence on its fate. Tooby et al. describe this process as involving nongenetic intergenerational regulatory systems that serve the purpose of "helping individual development along pathways better suited to the conditions it is likely to face in its life" (2003, p. 859). Such cross-generational continuity would, through some as yet poorly understood mechanism of transfer to the genome, ultimately lead to the particular cross-generational species-specific characteristics of a species. And, presumably, significant disruptions in this cross-generational pattern would either lead to species extinction or to the emergence of more effective adaptation strategies. Gottlieb (2002) in fact says much the same when he notes that the emergence of *neophenotypes* reflects alterations in a species-typical environment, especially early in life. These alterations in species-typical environments are likely caused by a physical or geographic change such as drastic climate or geographic disruptions or through the exploratory migration of species to novel environments which in turn prompt novel adaptation strategies:

What needs to happen to bring about evolution is the production of animals that live differently from their forebears. Living, differently, especially living in a different place, will subject the animal to new stresses, strains, and adaptations that will eventually alter their anatomy and physiology (without necessarily altering the genetic constitution of the changing population). The new situation will call forth previously untapped resources for anatomical and physiological change that are part of the species' already existing developmental adaptability. At some time further down the road it is possible the genetic makeup of the evolving population may change, but by the time that happens (if it does) the new behavioral, anatomical, and physiological changes will already be in place. The neophenotypic pathway for evolutionary change is thus seen as (a) an alteration of development leading to a significant change in behavior, followed by (b) a change in morphology, and eventually, possibly, (c) a change in the genetic composition of the population. (Gottlieb 1997, p. 150)

Such an adaptive responsive process as described by both Tooby and Cosmides and by Gottlieb might therefore be more likely to influence the evolution of species more likely to exhibit exploratory behavior such as humans and other primates than species whose behavior is rigid and stereotypic. For these species, a significant change in context would more likely lead to extinction. Further, because newer species make use of seemingly novel adaptations as well as earlier evolutionary adaptations 'borrowed' from other species, these evolutionary newer, higher order processes within a species (e.g., mechanisms regulated primarily by the cerebrum rather than the cerebellum in humans) might be even more open and responsive to adaptational pressures than mechanisms regulating more basic functions.

What then about all those modules? If predictable ontogenetic development is in fact a reflection of stable genetic, epigenetic, and contextual variables, as both EP and DST argue, then whatever modules are in fact evident at any point in the life span must be the product of these combined systems rather than the expression of any one. This certainly does not preclude an examination of the influence of any one of the three, but for such an examination to be valid, it must be done with the recognition of the contributions of the other two. So, fingers, toes, eyes, noses, ears, and maybe even partner preferences all reflect the combined influence of one's genotype, one's epigenetic regulatory mechanisms, and one's context. And if such behaviors appear universal across seemingly different contexts, maybe it is because the differences in those contexts are simply more apparent than real. The same would hold for behaviors labeled as innate. These too would reflect the same regulatory mechanisms although some of these mechanisms acting prenatally may be less obvious than others. Only the most ardent environmentalist should be upset, for example, by claims of neonates showing some rudimentary sense of quantity (Spelke 1998; Spelke and Kinzler 2009) or very young infants a rudimentary sense of object permanence (Baillargeon et al. 1985; Baillargeon 2002, 2008). In both cases, such competencies reflect a species-typical ability to detect some degree of discrepancy, clearly an essential key to survival and both competencies must reflect, from both theoretical perspectives, the influence of all three regulatory processes. And the presence of either very early in life does not necessarily presume the final adult form of either, although because all these developmental systems are best appreciated as regulating probabilities, it does mean that, all other things being equal, some developmental outcomes are more likely than others.

If adaptability is essential to both species survival and evolution, then the more adaptable the species, the more likely it is to prosper. But adaptability of a species, particularly humans and primates is not uniform across the life span. Adaptability, or plasticity if you will, is greater early in development, and, as such, understanding the development of young children, young primates, and chicks may provide the best means to fully appreciate the interplay of ontogeny and phylogeny. Bjorklund (2003, 2006; Grotuss et al. 2007; Causey et al. 2008) makes just this point in his discussion of evolutionary developmental psychology. Humans enter the world with many more neurons than will be the case by the adult years, that is, we are ready for virtually all likely eventualities. Neural pruning then occurs over the course of a life span as particular contexts strengthen some connections and weaken others. This selective pruning represents a trade-off between openness and more enhanced and efficient processing. In other words, we become better at what we do but at the price of being as open to new experiences. For Bjorklund, this ontogenetic pattern has significant phylogenetic implications. The more open or plastic the organism, the more able it is to adapt effectively to a species-atypical environment. Such an adaptation may lead to the expression of a novel phenotype. If this novel phenotype allows for increased survival and reproductive fitness, then it will spread within the population and ultimately serve as the material on which natural selection acts. And when are we best able to demonstrate such plasticity? When we are "chicks."

6

So What?

One of the major – if not *the* major – points of this book is that the nature–nurture debate continues to be fought along two battlefronts and that these two fronts represent different levels of analysis, that is, different metatheories. One metatheory reflects a belief that the life sciences, including the study of human development, can best be modeled using the same principles as those that work so well for the physical sciences, that is, a belief that complicated phenomena can best be understood as reflecting the cumulative effects of several antecedents each acting essentially independent of each other. The design and use of methodology allowing for the partitioning of variance into independent main effects is consistent with this worldview.

Scholars working within this reductionist metatheoretical tradition, such as those involved in the classic debate, typically reject the notion that there is a metatheoretical value reflected in their work. Rather, they argue that as empirical researchers (whether the focus is on nature or nurture), they are simply capturing what there is to be found with respect to the causes of human development and, further, that their research methods are neutral, that is, if the data show that it is possible to partition nature and nurture, it is so because it *is* possible to partition nature and nurture, not because the methodology artificially creates such a result. And finding evidence for such main effects, they report them, to both the scholarly community and to the general public. And so people picking up their morning papers (or iPads, as might be the case), read that scientists have found the God gene (Wade 2009), that bad driving might be genetic (McHughen et al. 2010), that political orientation might be genetic (Alford et al. 2005), and of course that the poor may not only be poorer than the rest of us but in some way might be genetically different as well (Rowe 2005).

Those scholars working in a systemic or nonreductionist metatheoretical tradition, one reflected in the new debate, deny the independence of antecedents, especially of nature and nurture. They argue that the physical sciences are not the most appropriate model for the life sciences, either in terms of theory or method. Instead, they argue that reducing complex, interdependent factors into independent antecedents is empirically not consistent with the data and makes about as much sense as trying to understand the properties of water through the independent study of hydrogen and oxygen; it's just all wet. This holistic, systemic approach does not lend itself well to the morning tabloid; however, these announcements to the general public are more likely found on a segment of National Public Radio or in the Sunday *New York Times.* They don't have sharp headlines because invariably the message of such articles is that "it depends." The one exception to this generalization are those evolutionary psychologists who focus almost exclusively on the adult expression of what are seen as adaptive brain modules regulating specific aspects of adult life, typically having something to do with sex differences between males and females. Although these scholars acknowledge the interactive manner in which adaptations are acquired over evolutionary time, by focusing exclusively on the adult expression of these adaptations, they omit consideration of the ontogenetic manner in which such adult behaviors might have come into being, that is, their arguments become very similar to those of the behavior geneticists.

One often overlooked implication of this debate about levels of analysis is that if a reductionist level of analysis is not in fact the most appropriate means to a full understanding of human development, then we confront the reality that not only are behavior genetic accounts called into question but equally so are classic environmental accounts. If genes don't make you do it, then the same has to be true about your mother. You can't have it one way and not the other if a reductionist level of analysis is rejected.

Unfortunately, however, the frequent conclusion of "it depends" offered at the systemic level of analysis does not make for very good sound bites, certainly not as good as, for example, "scientists discover the God gene." Even if there is a slow but steady shift within academic circles from reductionist to systemic thinking, the information typically provided from scholars working in the classic tradition to the media for the general public remains very much in the dualistic, reductionist tradition. And, for the most part, this information is more likely to focus on the nature side than on the nurture side of the debate. Why this is so is an interesting question to ask. It may simply be that most people tend to equate anything genetic with something permanent, and having a permanent answer for many is preferable to having a temporary answer, that is, an answer more likely coming from the nurture side of the classic debate. Advice to parents changes, genes don't, the argument might go.

So What About Nature?

So what is the nature message?[1] No doubt there are in fact many, but it is worth considering two that have received a lot of press over the years – the (un)role of parents in children's development and the causes of success and failure in a meritocratic society. The question of the role of parents reflects behavior genetics focus on accounting for variability, in this case the claim that once parenting skills reach a minimum threshold, all parents are "good enough." The issue of meritocracy touches on social class differences and the question of race in explaining one's status in a society. These two topics touch on the fundamental implications for policy and practice that a behavior genetic perspective offers.

Sandra Scarr's behavior genetic work discussed in Chapter 4 serves as the intellectual base for the behavior genetic argument that parents do not play as large a role in determining their children's development as common sense might lead one to believe. The particulars of the argument were discussed in the earlier chapter but, in short, the argument is that as children get older, shared environmental variance (i.e., parents) becomes less important in explaining differences between children and nonshared environmental variance (i.e., children evoking and eventually actively selecting environmental niches) becomes more important. This intentional behavior on the part of the child is seen as reflecting the increasingly dominant role of the child's genetics. In other words, it isn't that parents don't influence their children's development; it is rather that they primarily do so during children's early years, decreasingly thereafter. To Scarr, at least, such a conclusion allowed parents to enjoy their children more because they no longer needed to worry that anything and everything they do will or will not lead to a life of great fulfillment or utter depravity (for a contrasting position, one emphasizing the value of "tiger mothers," see Chua (2011)).

Although Scarr's argument about good enough parenting (Scarr 1996) was made several years ago, it still is a topic in popular media, thanks in large part to the work of Judith Harris in popularizing Scarr's ideas. In her original

[1] This "so what" discussion is intended to present the implications of each perspective as the perspective presents itself. Unlike earlier discussions of the four perspectives, this discussion is not meant to be critical or evaluative but rather is intended to illustrate the policy and practice implications of each view. A more critical analysis is also offered in the final chapter.

book *The Nurture Assumption* (Harris 1999) and more recently a revised 2011 edition, Harris argues that it is really the peer group rather than parents that is the most influential factor in determining a child's development. It isn't that parents are unimportant; it is just that they are less important. And because children pick their peers (but not their parents), these choices reflect children's interests, aptitudes, and temperament, that is, qualities seen by Harris and other behavior geneticists as largely under genetic control.

This issue of choosing ones peers, for Harris, is particularly significant because she believes that children in schools tend to align themselves with peers who are either "prolearning" or "antilearning." And perhaps echoing behavior genetic claims for a genetic base to socioeconomic class membership, she notes in an interview that

> The tendency of kids to split up spontaneously into subgroups also explains the uneven success rate of programs to put children from disadvantaged homes into private or parochial schools. The success of these programs hinges on numbers. If a classroom contains one or two kids from a different background, they assimilate and take the behaviors and attitudes of the others. But if there are five or six, they form a group of their own and retain the behaviors and attitudes they came with. (Harris, quoted in Lehrer 2009, p. 3)

Other interviews on her work as well as articles discussing the significance of it have appeared in many on-line and hard-copy magazines and newspapers including *Time* (Kingsbury 2009), the *New Yorker* (Gladwell 1998), www.shine.yahoo (Alphonse 2011), and, in England, the *Telegraph* (Paton 2007).

Harris's reference above to children from "different backgrounds" brings us to the second "so what" with respect to behavior genetic arguments, namely, the long-standing claim that there is a genetic basis to socioeconomic status and that efforts to reduce social inequality will actually have the opposite effect, leaving the genetically "less able" in a permanent underclass. Jensen makes the claim in his 1969 paper, but the more popular presentation of the issue came originally through the publication of Herrnstein's *IQ in the Meritocracy* (1973) followed by a 1994 publication titled *The Bell Curve* (Herrnstein 1994) and most recently in a book by Charles Murray (2008b) titled *Real Education: Four Simple Truths for Bringing America's Schools Back to Reality.* In all of these sources, the same argument is made simply, clearly, and, if you accept a behavior genetic perspective, logically compelling. As Murray describes it in his 2008 book, the argument is simply that (1) ability varies, (2) half of children are below average, (3) too many people go to college, and (4) America's future depends on how we educate the academically gifted. These "simple truths" are based on an argument that intelligence is a unitary construct, that it is distributed normally in the population (hence the bell curve with 50% above the mean and 50% below) and that where one falls on this distribution is largely due to genetics rather than to circumstance.

There are significant social, educational, and economic policy implications to such an interpretation of socioeconomic differences, but two that Murray has focused on most recently concern higher education. Murray's major concern with undergraduate education is that it is, at the same time, not rigorous enough for highly skilled students and too demanding for those students whose ability levels, Murray argues, should have never led them to attend college in the first place. This second, presumably marginally competent group, is in college because college has come to be seen as the exclusive road to material and social success, and so students who in fact are not competent enough to successfully complete college attempt to do so anyway (even in what Murray sees as a watered-down curriculum), fail, and often leave with substantial financial obligations but no degree. To Murray, the better alternative is more emphasis placed on skilled vocational training for such jobs as plumbers or electricians and in more focused and shorter college-level professional programs that prepare one to take a national certification exam in an area rather than lead to, in Murray's view, an expensive and essentially worthless bachelor's degree. By this approach, those who had the competence (i.e., genetic capacity) to do rigorous undergraduate- and graduate-level work would do so and those less competent would still be able to establish a meaningful adult life through the skilled trades or service professions. For Murray, such service professions include journalism, social work, public administration, business, and education. But, he argues, because we continue to insist that the only road to success is through college, more and more students who are not really "college material" enroll and fail, even with a watered-down curriculum. Murray notes that historically an IQ of 115 or higher was deemed to make someone "prime college material." Given Murray's belief in intelligence being normally distributed, people with an IQ of 115 or above comprise about 16% of the population. But because 28% of all adults have BAs, Murray concludes that the IQ required to get a degree these days is obviously a lot lower than 115 (Murray 2008a, 2008b).

The "cognitive stratification" that Murray argues will result from political policies based on the incorrect belief that social class ability differences primarily reflecting differences in opportunity, is, according to Murray, already very apparent. Murray dubs this group, in a *Washington Post* op-ed (Murray 2010) a "New Elite" The New Elite maintain their stratification by living in selective politically liberal enclaves (i.e., the East Coast and the West Coast), and then marrying each other, thus combining their large incomes and genius genes, and then producing offspring who get the benefits of both. And for Murray, an elite who passes only money to the next generation is evanescent. But an elite who also passes on ability is more tenacious, and the chasm between the elite and the rest of society widens (Murray 2010).

Murray does not offer an alternative to such elites, but presumably it would lie in his shifting more people away from traditional college programs into vocational and focused professional tracts, that is, tracts in which people of more "modest capacity" could be successful. For Murray and others who share his perspective, such social policies would presumably provide a broader range of people with job opportunities consistent with their particular level of abilities. And if, like Murray, you accept the notion that one's particular abilities are largely genetic and normally distributed, then his strategy certainly has a logic to it.

So What About Evolutionary Psychology?

So what about evolutionary psychology? Here too there are significant implications about policy and practice, but in comparison with the behavior geneticist perspective that tends to focus on socioeconomic status, the implications of an evolutionary psychological perspective tend to focus almost exclusively on gender differences.

In the concluding discussion on evolutionary psychology in the last chapter, I indicated that there really seem to be two evolutionary psychologies, one focusing on the ontogenetic implications of an evolutionary perspective and the other on the presumed genetically coded adult expression of the various adaptive cognitive and emotional modules acquired over evolutionary time. In the first case, the implications concern the value of immaturity. In the second, the implications concern how flexible our behavior is, especially our behavior as males and females, during the adult years. Each tract offers significant but different implications for both policy and practice.

An evolutionary perspective on ontogeny most closely reflects the work of Bjorklund (2006, 2007) but, as he notes, his conclusions are also consistent with progressive child development scholars including Neil Postman (1982) and David Elkind (1981, 1987). Neither Postman nor Elkind take an evolutionary perspective on child development, but their conclusions are similar to those of Bjorklund. The unique contribution of Bjorklund is situating this rationale within an evolutionary psychological perspective.

As discussed in Chapter 5, Bjorklund identifies two distinct sets of adaptations acquired over evolutionary time. One set he refers to as *deferred adaptations*. These adaptations are reflected in children's behavior that is seen as having the purpose of preparing children for their eventual roles as adults. For example, the fact that boys are more likely to engage in rough-and-tumble play whereas girls are more likely to show interest in caring for younger children is seen, by Bjorklund, as ways in which children begin to practice the evolutionary-based sex roles that they will come to assume as adult men and women.

But not all adaptations are seen by Bjorklund as deferred; others, *ontogenetic adaptations*, are seen as having an immediate value and actually have little, if anything, to do with future adult roles. These ontogenetic adaptations help children survive in the present, an especially important issue given the fact that our species has such an extended period of immaturity. The number of involuntary reflexes that infants demonstrate is a good example of this ontogenetic principle. In humans, the orienting, sucking, and startle reflexes help ensure that the very young infant will receive adequate nourishment and be less likely to fall. Other reflexes, such as the grasping reflex, might have little value to humans because parents hold on to their infants but nevertheless reflect our primate heritage because the young of other primates must be able to cling to their parents. In all cases, these reflexes eventually disappear as more flexible and adaptive strategies replace them. They have no adult-related role other than simply increasing the likelihood that the infant will some day actually become an adult. The same ontogenetic adaptations can be seen with respect to early attachment behaviors and wariness of strangers. These too have important survival values for the young but typically have significant long-term effects only when these early interactions are, for whatever reason, seriously compromised.

For human development to occur optimally, given its long period of immaturity, there must be ample opportunity for children to exercise both their deferred and ontogenetic adaptations. And herein lies the policy issue for Bjorklund, namely, the concern that efforts to accelerate children's development will leave less opportunity for children to, in effect, be children, and the less opportunity that children have to be children, the less likely they are to become effective, mature adults. And of all the ways in which society seems to be conspiring to rob children of their childhoods, Bjorklund sees the assault on play as particularly harmful:

> Play is the natural expression of youth and the way in which children interact with the world. Play is a reflection of children's developing competencies, but in turn facilitates the development of those, and other, competencies. Play is the joyful conduit by which experience changes children's brains and minds. Children may not set out to learn, or discover, anything via play, but they do anyway. Learning

can be more explicit, or intentional, as often happens in formal schooling, but parents and educators should not ignore how much children learn through play, usually without even being aware of it. (Bjorklund 2007, p. 162)

The problem, for Bjorklund, is that when play-based or discovery learning is replaced by some form of adult-directed academic instruction, young children find themselves being asked to behave in ways that are inconsistent with what both ontogeny and evolution have designed for them. Such academic strategies do lead to specific improvements in very narrowly defined areas but do so at the cost of robbing children of more appropriate opportunities to be engaged in both cognitive and social pursuits. Further, these academic gains are often short lived and can introduce a good deal of unnecessary stress into the lives of young children and their families who have come to believe that sooner is better even though the research evidence does not support such a position. For Bjorklund, the policy implication is not that early education programs are inappropriate for young children but rather that these programs should be consistent with the developmental nature of young children's evolutionary-based efforts to make sense out of their early experiences.

As concerned as Bjorklund is about the loss of playfulness in the lives of young children, he is perhaps even more concerned about the loss of childhood for children between the ages of 7 and 12 years. These children are exposed to increasing academic pressures, expectations to dress and behave in ways more appropriate (or inappropriate) to adolescents or even adults, and are often left to their own devices as parents find themselves increasingly involved in their own adult responsibilities. The issue is that because modern life is so complicated and increasingly demands high levels of both social and cognitive competence, middle childhood should be a time of providing children the experiences necessary to foster greater plasticity in their thinking rather than a time of increasing focus on excellence in a few areas. Again, if we don't provide time for children to be children, then, from an evolutionary perspective, we are making it just that more difficult for adults to be adults:

Modern life is enormously complicated. We may not need to learn how to hunt gazelle, forage for nutritious tubers, or start a fire with flint: but we do need to learn how to read, to calculate, to operate modern machinery, to drive and to buy plane tickets on-line. We need to learn how to negotiate with one another, to avoid fights, to woo members of the opposite sex, and to get along with our neighbors and bosses, who may very likely be from different tribes than we are. It is because we have so much to learn that we need an extended period of immaturity to learn it. (Bjorklund 2007, p. 217)

For Bjorklund and others who lament the loss of childhood, the policy implications are clear. The longer into the life span we can foster a sense of intellectual and social playfulness in our approach to life's challenges, the better able we become to continually find new ways to adapt to an increasingly changing world. Said another way, efforts to "accelerate" learning and development, particularly through the type of high-pressure, intentional, didactic, adult-directed instructional strategies described in a recent *New York Times* article (Zernike 2011) are not only likely harmful to children but certainly have long-term effects as well because they interfere with the opportunities children need to practice evolutionary-based adaptations. Our brains may not have changed much over the past thousands of years. But what evolution has provided us is a brain that can remain open to new ways of doing things as long as our environments continue to provide us opportunities to learn new things in ways consistent with our evolutionary heritage, that is, opportunities that foster a prolonged period of immaturity.

Whether or not we are fortunate enough to enjoy a prolonged period of immaturity in our development, eventually we do find ourselves dealing with the issues of adulthood, and here too evolutionary psychology proposes several insights for how best to do so in ways consistent with our evolutionary heritage. Perhaps the most basic concerns the argument that there is a fundamental human nature that transcends cultural differences. Contrary to the arguments of some anthropologists and postmodern theorists, evolutionists argue that all cultural practices reflect a shared evolutionary history based on a content-structuring, evolved organization of the human mind. "Evoked culture arises from human cognitive architecture and expresses itself differentially according to local conditions" (Confer et al. 2010). In other words, all cultures must deal with the same evolutionary-based issues such as organization and hierarchy, establishing and maintaining relationships, conflict, the acquisition of resources, and the transmission of cultural values and beliefs across generations. These are universals, they reflect who we are as a species. The variations that are seen across cultures, even when they are so distinct as to seem solely unique, are, according to the evolutionists, simply the interaction of our shared human nature with the particulars of time and place. You simply don't see cultures that have not structured themselves along these evolutionary-related dimensions. Simply put, for evolutionists, we are much more like each other than we are different from each other, a perspective with very significant implications for social policy and practice.

Although we may all share a common human nature, we nevertheless reflect two distinct sexes, and, for evolutionists, these sex-based differences permeate virtually every aspect of our lives, reflecting the ramifications of the distinct roles that males and females play in reproduction, the bottom line for any evolutionary analysis. So evolutionists note significant differences in both the short- and long-term mating strategies of males and females, in their respective preferences for erotic materials, in differing definitions of sexual harassment, and in parenting behavior, to name just a few. And, of particular relevance for the present discussion, each of these reported differences has implications for the way that a society structures itself and in the ways that this society creates laws related to such issues.

Consider the issue of sexual harassment. The legal standard of what constitutes harassment is "the reasonable person standard," that is, what most people would describe as behavior that is either fear inducing or harassing. But, as Duntley and Buss (2010) argue, an evolutionary perspective puts a new light on the reasonable-person hypothesis. They argue that evolution-guided research shows that women consistently judge a variety of acts to be more sexually harassing than do men, and that women experience greater levels of fear than do men in response to specific acts of being stalked. As such, an "ungendered" reasonable-person standard, the legal criteria for harassment, may be an inappropriate criterion.

Evolutionists are quick to point out that research on topics such as harassment and, even more so, on rape (Thornhill and Palmer 2004) is not meant to imply that either harassment or especially rape is an acceptable behavior (a charge often leveled against evolutionists). Rather, the argument is always that if society wishes to rid itself of such offensive acts, then the only effective way to do so is to have a true understanding of their origins.

This same perspective of understanding origins applies equally to one of the most basic of our social values, monogamy in marriage. Writing from an evolutionary perspective, Ryan and Jetha (2010) argue that we were never meant to be a monogamous species and in fact that monogamy, as a socially valued concept, became recognized only when our species first made the transition from a hunter–gatherer culture to an agrarian culture, a mere 10,000 or so years ago. Being able to stay in one place meant that, for the first time, we became able to amass possessions because we no longer needed to worry about traveling light. Hunter–gatherers are said to have shared everything, including each other, as an adaptive survival strategy. But once we started gaining possessions, we became more protective of what we had, including our partners; hence the origins of the value of monogamy. Coveting anything of thy neighbors now became a bad thing to do.

So, should we be surprised that the divorce rate is so low rather than so high? Should we no longer be surprised when so many of our public figures, so-called "alpha males," are found to have engaged in extramarital affairs? What are the policy implications? Should we simply become more tolerant of people "sleeping around?" Or should we redouble moral, legal, and even economic efforts to support fidelity? Again, the evolutionists would argue that showing us not to be a "naturally" monogamous species does not directly translate into any specific policy recommendation. But, again, they would argue, it does highlight the necessity of recognizing our nature if we wish to change that nature.

One final example of an evolutionary take on public policy and practice, this time in the workplace. One of the most common policy issues we deal with as a society is that of gender equity in the workplace. We know that there are significant differences in the number of males and females entering individual occupations, that there have been and continue to be significant salary differences between men and women, and that as one examines corporate ladders, one increasingly finds them to be dominated by males. Such observations are frequently the source of class-action lawsuits alleging gender bias, of efforts to equalize the percentage of males and females in particular occupations, especially math, science, and engineering, and legislative efforts to remove what are seen as institutional barriers to opportunities for both equal pay and equal advancement. The presumption in all of these efforts to remedy this gender bias is that they all reflect the fact that men have enjoyed positions of dominance in the workplace and, as such, have instituted practices, intentionally or unintentionally, that serve to ensure the continuity of their privileged positions in the workplace. In other words, in a society of true free choice, in which the work options for men and women were truly equally available, in which there was no bias, the argument is that we would indeed expect to find gender equity in terms of occupational choice, salary, and participation in the boardroom. The evolutionists see it a little differently.

Browne (2004) argues that such equity claims are based on the implicit assumption that male and female brains are sexually monomorphic and, as such, process information in the same ways. But, the evolutionists argue, male and female brains have evolved in different ways, largely as a reflection of their distinct and specific roles with respect to reproduction and parental investment. And because these evolutionary-based sex differences are seen as having consequences for the behavior of males and females in a variety of settings and relationships, this should equally be the case with respect to the workplace:

> Any account of women in the workplace must explain, not imply, why, on some global measure, women have not advanced as far as men. A low-resolution view of the workplace might make such commonly invoked causes as discrimination

by employers and a generalized sexism on the part of society plausible candidates. A higher-resolution view of workplace patterns, however, reveals a rich texture that makes simplistic explanations unconvincing.... A more nuanced analysis of the workplace reveals that the variegated patterns are more readily explainable in terms of evolved sex differences than they are by some generalized implied conspiracy against female achievement. Average sex differences in interests and abilities can explain not only why women have failed to make inroads in certain areas, but also why they have made such stunning advances in others. (Browne 2004, pp. 278–9)

So, Browne argues, the greater percentage of men at the tops of corporate ladders are more likely a reflection of the evolutionary pattern of men being more competitive and more interested in both status and dominance, largely a reflection of their evolutionary role in competing for female partners. Women, he argues, seem to have less evolutionary investment in status and dominance and greater investment in parenting. As such, women are less likely to demonstrate the "single-mindedness" that Browne argues is the essential element in making one's way up the corporate or entrepreneurial ladder. The fact that women are significantly less represented in high-risk professions (both in terms of physical risk such as firefighter or in financial risk such as stockbroker) again reflects the greater evolutionary payoff that men are seen as having in evolutionary terms, that is, the most preferred partners.

Browne argues that because there are important physical and cognitive differences between men and women, we should not be surprised to find more men in both physically demanding jobs and in those required excellent spatial, mathematical, and mechanical skills. Browne is not suggesting that women should not be firefighters or engineers, only that our evolutionary heritage means that they are much less likely to pursue such roles, even if there truly was equal opportunity. In general, Browne argues, women prefer jobs with safe working conditions, regular or flexible hours, pleasant surroundings, short commutes, and good relations with co-workers and supervisors. Men, on the other hand, are seen as focusing solely on salary and promotion opportunities.

Like more evolutionary psychologists, Browne goes to some length to caution against making the naturalistic fallacy, that is, mistaking "is" for "ought." Instead he makes two points. First, we should recognize that in some cases, differential employment patterns may in fact reflect the exercise of free choice for both men and women rather than some form of bias. If evolutionary pressures have led to differences between men and women, then it is certainly reasonable to expect that today these differences would be reflected in employment patterns. Second, to the degree that bias does exist in hiring and promotion decisions, then the only way to truly rid the workplace of them is establish legal policies and practices that our consistent with our evolutionary heritage. "Nonetheless, evolutionary psychology offers policymakers a window into the fundamental nature of the human animal and, for that reason, holds out the hope of policies that are more effective than those formed in studied ignorance of human nature." (Browne 2004, p. 290)

So What About Nurture?

So what about the implications of a nurture position? First, unlike the implications that evolve from a behavior genetic or evolutionary psychological perspective, these will be much less specific, both because the environment is typically depicted as potentially more open-ended and because individual environmental events are typically not seen as exerting lasting, significant developmental effects. Rather, it is their cumulative impact that is potentially significant. Thus a discussion related to policy and practice would need to focus on the value of cumulative, sustained settings or interventions as the most appropriate means through which to influence the course of development.

One only needs to reflect on Evans and English's (Evans and English 2002; Evans 2004) findings concerning the impact of socioeconomic status on development and on what it would take to address these multiple issues to appreciate the cumulative of individual environmental events. Crumbling schools, inadequate housing, limited access to health care, and exposure to pollutants exert their pernicious influence over periods of time, and it will likely take at least as long to remedy the situation as it has to cause it. Further, effective remedies will not occur if the focus is just on schooling or just on housing or just on health care or just on pollution. The nurture perspective makes clear that it is the cumulative impact of all of these disadvantages that is the issue, not any one in particular. By the same token, it is equally important to recognize that a structurally similar set of environmental forces serves to foster the development of those in more advantaged circumstances. And for those who are advantaged, a negative shift in one environmental element is no more likely to have a prolonged, seriously adverse effect on development than the introduction of, for example, decent housing alone will offer a significant advantage in the lives of those less privileged. Rutter et al. (2001) make this point very clearly when they talk about the curvilinear pattern of influence when additional stressors become present in people's lives. Greater stress places a disproportionately greater burden on individuals.

Second, because the environment is depicted as existing at multiple levels, the implications that follow would also tend to focus on what policy and practice recommendations would most appropriately be made with respect to activity at each of Bronfenbrenner's four levels of the social ecosystem. His work (Bronfenbrenner and Evans 2000), as well as that of Kohn (2006) and Elder (1998), suggest that for both theoretical and practical reasons, interventions focusing on social address variables at the level of the exosystem may offer potentially better long-term effects than interventions at more immediate levels. Bronfenbrenner, for example, notes the significance of "chaotic systems" on development. He describes such systems as involving frenetic activity, lacking social structure, and therefore highly unpredictable from day to day and involving a high level of "ambient stimulation." Such chaotic systems interfere with the proximal processes that engender competence but instead produce proximal processes that in and of themselves lead to dysfunction. Kohn's emphasis on characteristics of the workplace and Elder's emphasis on exosystem factors such as economic and political climates reinforce Bronfenbrenner's arguments and, taken together, appear to make a clear argument, from a nurture perspective, for broad, exosystem social address interventions.

Third, because there is not one but potentially many outcomes resulting from sustained, cumulative environmental events, evaluating the efficacy of any intervention requires attention to multiple outcomes, some expressing themselves in the short term and some in the long term. Looking at only one outcome such as change in grade point average or at only one time of measurement will most likely understate the potential impact of environmental events. One excellent example of this policy and practice implication is the long-term follow-up research on early childhood intervention programs.

There are now several excellent studies (Calman and Tarr-Whelan 2005; Schweinhart et al. 2005; Barnett 2008) looking at the long-term impact of publicly supported early intervention programs. These programs have their origins in the 1960s as part of the then War on Poverty and continue today through Project Head Start at the federal level as well as many programs funded at the state and local levels.

Of these many studies, probably the most well known and best designed is the HighScope Perry Preschool Project established by David Weikart in 1962 (Weikart et al. 1970). The Project initially involved 123 children from low-income households who were assessed to be at high risk for school failure. By random assignment, half of the children participated in the Project, the other half did not. The children have continued to be followed as they moved through the K-12 school system and during their adult years. Across this entire age range, there remain striking differences between the two groups of children. At age 5, at the conclusion of the Project itself, twice as many children had IQ scores above 90; by age 14 the HighScope participants were more than three times as likely to be at grade level in school; 77% graduated high school compared with 60% for the control group; and by age 40, the HighScope children were more likely to be earning a wage of at least $20,000 and almost half as likely to have a significant arrest record defined by five or more arrests. To put these differences in economic terms, the cost to provide services to each High/Scope participant was just over $15,000. The economic benefit derived from participation in the Project was estimated to be just under $200,000. This estimate reflected the fact that the HighScope children were less likely to receive costly special education services while in school, that they were less likely to receive public assistance during the adult years, that they were less likely to be incarcerated, and that they were more likely to be tax payers.

For the present discussion, there are two messages to take away from this data, both pointing out how best to conceptualize the impact of environmental events on development. The first concerns the issue of continuity and the second, the issue of specific outcomes over time.

No one, including the HighScope researchers, suggests that their 2-year early education program somehow "inoculated" the children against future risk and trauma. That strategy may work for the measles but it doesn't seem to work for developmental issues. Rather, the HighScope researchers posit a path model that shows that children who are more intellectually competent when they enter kindergarten have better educational and interpersonal experiences in the primary grades, which increases the probability that these early successes will compound over the rest of their school years, which then increases the likelihood that they will successfully enter the workforce, be able to hold on to a job, make a decent living, and therefore be less likely to either need public assistance or find themselves dealing with the criminal court system. The relationship is not perfect by any means; 60% of the High/Scope children did not graduate high school and 36% were arrested five or more times by the age of 40. But as depressing as these outcomes are, they are clearly an improvement over the life histories of the children not enrolled in the HighScope Project. So, as Bronfenbrenner makes clear, it is only when an initial advantage is maintained over a significant period of time through an appropriately responsive social ecosystem that we should expect to find the maintenance of this initial advantage over that extended period of time. In other words, it isn't just that the HighScope children were different after 2 years than the control children, it was also that an appropriately responsive environment reacted to the HighScope children in ways different from how it reacted to the control children. Without this sustained nurturance, the initial advantage would have quickly faded. It is reasonable to argue, from a nurture perspective, that if the follow-up environments of the HighScope children had been even more responsive to these initial differences then the differences over time would have been just that much greater.

The second implication coming from the HighScope data concerns the question of how best to measure the effect of the environment over time. The HighScope data as well as other reviews of early interventions report a variety of outcomes, most correlated with the age of the individuals. There is certainly a logic to the pattern of findings reported – early academic success predicts later work and social outcomes – but there is also a post hoc quality to it all. It is unlikely that the HighScope researchers or any other early interventionists for that matter would be able to identify at the start of a program for preschoolers what the likely outcomes would be by the time the children were 40 years old. Again the issue is one of there not being an effective and comprehensive model of the impact of the environment, in part because we still lack a good conceptual model of how nurture works over time and because, when we do have such model, it will almost certainly be probabilistic rather than deterministic. The HighScope gains were not guaranteed to last and did so only because the probability of subsequent positive interactions was greater for the HighScope children than for the control children, but had the economic collapse of the 2000s occurred in the 1990s, the long-term data would most likely have shown little to any lasting effect.

Fourth, consistent with Elder's life-course analysis, recommendations for policy and practice based on environmental data must always be situated within a sociohistorical context because the interplay of factors across cohorts often differs. As Elder found, the impact of the same set of social and economic factors – the Great Depression of the 1930s and World War II – depended on a person's particular birth cohort.

And finally, the impact of environmental change takes time, sometimes a long time, even at times spanning generations. Perhaps nothing better illustrates this last point than the conclusion Sarason (1973) draws from his personal narrative of growing up Jewish:

> I have no difficulty accepting the notion that intelligence has its genetic components, not do I have difficulty with the idea that different groups may possess different patterns of abilities. It would require mental derangement of the most serious sort to deny that different groups get different scores on various tests

of intelligence. But I have the greatest difficulty understanding how anyone can come to a definitive conclusion in these matters based on studies which assume that what culture and history have created can be changed in a matter of years or decades. What combination of ignorance and presumption, what kind of understanding of human history does one have to possess to accept the *hypothesis* that the central psychological core of *historically* rooted groups can markedly change in a lifetime? [Italics in original.] (p. 967)

So What About Developmental Systems Theory?

So what are the practical implications of a developmental systems perspective? In their response to Lickliter and Honeycutt's (Lickliter and Honeycutt 2003) critique of evolutionary psychology, Tooby and colleagues (Tooby et al. 2003) took the opportunity to lob a few shots back at developmental systems theory (DST). One of the arguments they made was that DST has no utility because it offers no specific predictions about behavior or development. Well, they were certainly correct in arguing that in comparison with the many, highly specific predictions about virtually all aspects of development offered be evolutionary psychology, DST does not offer anything comparable. But they were wrong when they claimed that DST has nothing practical to offer at all because it has this to offer: *it depends.* This may not seem like much compared with the often highly specific predictions made by the three other perspectives but in fact it is just as valuable a contribution because it offers a counterbalance to the "one-size-fits-all" main-effects message commonly offered by the other three. Said another way, the other three perspectives tend to offer policy and practice recommendations at a macrolevel; "it depends" suggests that a microlevel approach might be more appropriate.

"It depends" itself depends, reflecting the probabilistic nature of DST. Because these probabilities depend on particular genotype–environment interdependencies, some outcomes are seen as more likely than others and no outcome is seen as universal. On the one hand, we can be pretty confident about the number of fingers and toes we come into the world with because there is relatively little variability in the developmental systems regulating limb formation. On the other hand, because there is much more variability present in more interpersonal domains such as parent–child relationships, predictability is less accurate. Theoretically, such a probabilistic perspective might be seen as arguing that anything is possible but, in fact, even though DST recognizes organisms as much more open systems than, for example, evolutionary psychology does, it does not claim that anything is possible because the system is bounded by the synergistic expression of its components (Tudge et al. 1996). This is of course very much the same line of reasoning that Anastasi offered many years ago (1958).

So what then does DST really have to offer with respect to applied policy and practice? If one-size-fits-all main-effects strategies are not appropriate, what is? How is it possible to develop effective applied policy and practice if the most you seem to be able to guarantee is that "it depends"? The answer is that you need to look more carefully on what "it' in fact does depend on.

DST as an integrated approach makes clear that multiple factors interact in a synergistic fashion both in terms of the development of behavior and its expression. As such, the first applied recommendation of a systems approach is that all policy and practice planning must reflect input from the various disciplines believed to have some influence on the outcome behavior or behaviors. Rutter makes this point very clearly when he talks about trying to understand why psychological disorders have increased so dramatically in young people over the past 50 years:

> Why has this been so? I would argue that this has to be an answerable question. If we had a proper understanding of why society has been so spectacularly successful in making things psychologically worse for children and young adults, we might have a better idea as to how we can make things better in the future. To succeed in this gargantuan task, use of a diverse range of research strategies is necessary. The answer will not come from genetic research on its own, or from environmental studies, or developmental investigations; the combination of the three might do much, however. It will be necessary to recognize the range of different causal questions to be considered. The explanation for individual differences may not be the same as that for differences in the level of the trait or the frequency of the disorder. (2002, p. 15)

In other words, all planning and implementation of policy and practice must be done in a collaborative fashion involving representatives from all relevant perspectives. This isn't exactly a particularly earth-shattering implication except for the fact that, as often as not, it doesn't happen. Just talk to a parent of a child with significant, multiple developmental challenges about how well the child's different specialists coordinate their services. Our medical and social service delivery systems have historically been highly bureaucratic and often adversarial with respect to each other, each privileging its particular "organ" or service above all else. DST implies that such bureaucratic systems can be both ineffective and inefficient because, as Magnusson and Cairns (1996) note, individuals function as integrated organisms: "Single aspects do not develop and function in isolation, and they should not be divorced from the totality in analysis" (p. 12).

The second implication of DST is that assuming these diverse perspectives are able to work collaboratively, their recommendations need to allow significant flexibility so that all those affected can be appropriately accommodated. No single approach to instruction works for all children; nor does any one single approach to evaluation. Even when we are able to show a significant advantage for one instructional approach over another, it is still almost always the case that a closer inspection of the data reveals that in spite of the fact that the group mean of one group was statistically higher than that of the other, some in the higher group will have nevertheless still done poorer than some in the lower group. And yet we continue to favor educational interventions based on aggregate data and then are surprised when the results do not show consistent, uniform benefits over time. Instead of standardizing the evaluation and seeing what variability results in the outcomes, as might be the approach of behavior genetics, a DST perspective implies that we should be standardizing the result and varying the input necessary to achieve it in each individual (Lerner 2006).

Kagan (2006) reminds us of a third implication of DST, namely that the study and fostering of development must be done at a holistic level because attempts to reduce psychological phenomena to either their genetic or environmental antecedents eliminates the very things we wish to study and foster. Kagan's particular concern was with those who argue for a close correspondence between brain locations and psychological activity, especially those who claim that the identification of such a brain location is also an identification of the corresponding psychological action. He argues that there are no such reliable correspondences, that the brain tends to react differently to the same exposure if the context of that exposure is different, and that it is the interplay between structures that reflects psychological meaning rather than that meaning residing in any one structure. As such, the effectiveness of any effort to influence some aspect of development can be measured at only the psychological level, even though inputs to the outcome behavior reflect activity at all levels of the organism.

So, So What?

The theoretical issues that have been discussed throughout this book now come to play an even larger role when the focus shifts from theory to practice. It is one thing to debate the significance of our evolutionary history; it is another thing to perhaps suggest that the reasonable-person standard has no legal validity. It is one thing to argue the value of educational standards; it is another thing to expect all children to respond equally well to high-stakes standardized testing. As has been the case throughout this discussion, the issues are ones of modifiability and uniformity. How much can people change? What is the best way to do so?

Not surprisingly, each of the four perspectives offers a different answer to each of these two questions, but what is also interesting to note are the differences in the number and degree of specificity of the practical implications offered by each. Clearly, the nature perspective and the adult-focused predictions of evolutionary psychology proffer more than the other perspectives. This does not necessarily mean that these two perspectives are of more utility; it may just mean that these researchers are "bolder' or that their models lend themselves more to specific predictions. Both of these latter two explanations are likely because both perspectives make greater use of predictions as a way to further establish their legitimacy as explanations for development, and each offers a unit of analysis (genes in the first case; modules in the second) that is touted as having a virtual one-to-one correspondence with specific developmental outcomes. The nurture perspective also allows for specific predictions but because it is somewhat hobbled by a lack of a standard unit of measurement, its predictions are always couched within the confines of the particular intervention or research agenda. And then there is DST with its message of caution. Our political and social systems are organized in ways more consistent with the broad strokes of the first three perspectives, but it would be interesting to speculate what these systems would look like if they were more consistent with a DST perspective. So who is right? Maybe it just depends.

7

Now What?

There is no question but that one's genotype significantly influences one's course of development. And there is no question but that one's environment significantly influences one's course of development. But there is now also no question that the way that the classic debate has attempted, for more than 100 years, to explain the role of each has done more harm than good, both in terms of our conceptual understanding of the course of development and in terms of the inappropriate policy and practice recommendations that have flowed from such a flawed perspective. As I said earlier, it really is incredible.

The classic debate's reliance on main-effects statistical models simply no longer reflects what we understand about the interdependence of nature and nurture. A statistical model not only needs to be mathematically accurate; it needs to accurately model the phenomenon of interest as well. The main-effects model may pass the first test, but it no longer passes the second test. In fact, it may never have passed the second test because partitioning variance is at heart an illusion. There is in reality no such thing as an independent effect.

Think of it this way. As all of us remember from sixth grade algebra, an equation with the same terms on both sides may be simplified by canceling or deleting the two identical terms. Doing so does not affect the value of the equation. Even preschoolers have an at least an intuitive understanding of this principle when they add and remove weights from a balance beam. This mathematic truth serves as the rationale for partitioning variance. By averaging across one independent variable, the influence of that variable on the dependent variable is presumably canceled. In other words, in, for example, a simple 2×2 design with two levels of each independent variable, the strategy for finding the percentage of variance accounted for by independent variable *A* is to average across both levels of independent variable *B*. The argument then becomes that because the influence of variable *B* is "counterbalanced" across both levels of variable *A*, then the only possible explanation for a significant difference between the two levels of variable *A* in terms of the dependent measure must alone be variable *A*. The same logic would hold for looking at the variance attributable to independent variable *B*; you would average across independent variable *A*. Mathematically speaking, all fine and good.

But here's the rub. Averaging across a variable doesn't eliminate its influence, and it counterbalances that influence only if the two variables are, in fact, truly independent. But if antecedents are in fact interdependent rather than independent, then nothing has been controlled, a point made by others as well (Johnston 1987; Wahlsten 1990; Vreeke 2000; Lewontin 2006; Rutter 2007). All this might not be an issue – as the statisticians like to say, the procedure might be "robust" – if these interdependences all acted in the same manner and direction, but the more we come to understand about development, the less robust such claims become. Is it therefore any wonder that within-cell variance is typically high even in a well-designed study and why, ironically, the findings seem to ultimately account for so little variance.

Reductionist, main-effects, Cartesian dualistic models of human development simply do not provide an accurate image of the process of development. I am hardly the first to make this claim, and unfortunately I probably won't be the last, but I gladly add my name to the list of those who over the years have made the same argument (Carmichael 1925; Anastasi 1958; McCall 1981; Ehrlich and Feldman 2003; Gottlieb 2007; Meaney 2010). It is now simply time to move on.

Okay then, now what?

Piagetians note that children typically begin to recognize values as existing along continuous dimensions between the ages of 5 and 7 years. Before this point, a preschooler might simply recognize one person as tall and another as short without realizing that height, including the heights of the two people in question, can exist along a continuous dimension that, in this case, is called height. What was once seen as one person as tall and another as short is now recognized as one person being *taller* than the other. Although children seem able to begin this transition to dimensional thought early in life, it appears to be taking considerably longer for developmentalists to reach the same milestone (Stotz 2008). Nature and nurture are just like tall and short. But hope springs eternal, and there is now a growing body of evidence that suggests that even developmentalists are coming to be able to think dimensionally.

This last chapter looks at this newly "emerging competence" of developmentalists. I first discuss progress in both the disciplines of biology and human development that have fostered this dimensional (actually multidimensional) thinking and then propose an integration of the two protagonists in the new debate, namely, developmental systems theory (DST) and evolutionary psychology (EP). It won't be a perfect fit but, as Pepper (1961) concluded in his analysis of worldviews, it will be "reasonably adequate" and certainly a starting place for the development of a *new synthesis.* Finally, I discuss the implications of an integrated "new view" for the study of development and, by implication, the preparation of future developmentalists. Let's start with the biologists and their revision of the *modern synthesis.*

Evo–Devo

The modern synthesis in fact isn't that modern. It became codified in the 1930s and 1940s as research at that time led to an integration of natural selection as the driving force of evolution with genetic theories of inheritance (West-Eberhard 1998; Bjorklund 2006; Lickliter 2008). Evolution was now seen as occurring through small, genetically mediated changes in the genome and because cross-generational gene inheritance was considered the only mechanism through which species specific characteristics could be passed and maintained across generations, what occurred after conception, that is, ontogeny, took a back seat to phylogeny. It wasn't that what developmental biologists (*née* embryologists) did was unimportant; it was just, according to the proponents of the modern synthesis, that what they did didn't have anything to do with the evolution of a species, or, as Bjorklund (2006) puts it, "development was an epiphenomenon of evolution; it may have great consequences for the individual, but it is inconsequential for phylogeny" (p. 214). The modern synthesis simply recognized no mechanism through which events occurring during ontogeny (with the exception of genetic mutation) could influence phylogeny. And, of course, this effective separation of nature and nurture within biology was echoed by psychologists favoring the role of instincts (Lorenz 1965) in development as well as maturational accounts of development (Gesell and Ilg 1949; Lorenz 1965). But that was then, and continuing research on genetic mechanisms has led to a reevaluation of the role of genetics in development and, as a result, a reexamination of the relationship of phylogeny and ontogeny.

The newly emerging perspective, evolutionary developmental biology or *evo–devo* for short, has identified a number of nongenetic or *epigenetic* mechanisms regulating the developmental process both within and across generations and, as a result, is increasingly putting focus on the possible role of ontogeny in phylogeny. In particular, a focus on epigenetic mechanisms recognizes the fact that development is not precoded in the genome but rather reflects the continuing bidirectional interactions of variables across all levels from the molecular to the broadest social environment. It is in this sense that Gottlieb (2007) referred to development as probabilistic rather than predetermined and, in a similar vein, why the evolutionary psychologists (Tooby and Cosmides 2005) also recognize the equally essential role of environment in the evolutionary process.

As initially discussed in Chapter 5, epigenetic mechanisms serve as the interface between the genome and its environments and as such serves as a regulator of genetic expression. One consequence of the increasing understanding of epigenetic mechanisms is that the role of the gene itself has shifted from being seen as the determinant of development to one seen as the provider of essential data to the developmental process (Falk 2000). As Lewkowicz (2011) describes this redefined process:

> Gene expression does not involve the simple read-out of the linearly arranged set of nucleotide base pairs. Instead, it is controlled by a cascade of factors all of which interact with one another in a sequential fashion. Specifically, transcription requires the action of two regions of DNA that are adjacent to the gene. These regions, the promoter and the enhancer, become activated by transcription proteins floating in the cell. Without the action of these proteins and the promoter and enhancer regions, no transcription occurs. These proteins, in turn, are regulated by signals from outside the cell that can be as far removed as the behavior of others. (p. 6)

One significant implication of this epigenetic view is that a full understanding of development, both when it proceeds as expected and when it doesn't, requires not simply a knowledge of the genetic code but equally so a knowledge of those factors likely to interact with DNA as well as the timing of these likely interactions. Further, these "nongenetic" factors should not be conceptualized as "background" factors because "they co-specify the linear sequence of the gene product together with the target DNA sequence" (Stotz 2008, p. 367). And, again, because such knowledge is not predetermined, development is therefore best recognized as a probabilistic process.

The presumption that development is a probabilistic process implies that there must be a degree of plasticity or variability that defines the developing organism, that the organism can change or adapt to changing conditions both in terms of the generation of novel responses as well as the maintenance of basic functions. So, on the one hand, we can identify such seemingly highly canalized functions as the "proper" number of fingers and toes or the individual's ability to maintain constant body temperature in response to changing environmental conditions, and, on the other hand, we can identify the expression of novel responses, as is often the case when considering higher-order cognitive functioning (Jablonka 2007; Lamm and Jablonka 2008). Both cases – maintaining a steady state and the expression of novel responses – reflect the organism's ability to change in response to context. To one degree or another, all species demonstrate this property, but the more complex the species, the greater the evidence of this plasticity. This degree of plasticity is further enhanced by the fact that epigenetic mechanism are potentially reversible (Szyf et al. 2008). Further, the effect is seen as bidirectional: Changes to the organism originating from the social environment that are then mediated through epigenetic mechanisms are reflected in the behavior of the organism, and this behavior, in turn, exerts an influence on that environment, further influencing potential epigenetic consequences. This was certainly the case with respect to the rat pups exposed to different maternal handling conditions reported by Meaney and Szyf (2005) as well as the studies of children by Caspi et al. (2002). For example, in Meaney's research, differential handling led to a set of epigenetic changes in the pups that became most evident in terms of stress responses that then in turn affected parental relationships with their own offspring, that is, the "grandpups" of the differentially handled rats.

As Meaney and others (Caspi et al. 2002; Meaney and Szyf 2005) have demonstrated, epigenetic mechanisms not only influence individual development but also demonstrate significant cross-generational effects as well. Evidence of these cross-generational effects is important for two reasons. First, it establishes a nongenetic inheritance mechanism, and second, it suggests an intriguing new perspective on species evolution.

The integration of ontogeny with phylogeny, one of the cornerstones of evo–devo, means that rather than natural selection acting directly on the genotype, it is more reasonable to argue that it acts on the phenotype. And because phenotype reflects an organism's efforts to adapt to a multilevel, multidimensional environment, especially in the case of more complex organisms such as humans and other primates, it is ontogeny that presents the behaviors on which natural selection acts. Such a perspective does not deny the important role of genetic mutation in the evolutionary process (or population drift or migration) but, as Jablonka (2007) notes, "the chance that a newly induced behavior modification can be positively selected is therefore substantially greater than that of a rare, mutationally induced phenotypic modification that occurs in a single individual" (p. 813).

Within any species population, prenatal and postnatal variabilities in an individual's developmental patterns in both the rate and timing of developmental experiences result in a degree of variability across the members of that population (Lickliter 2008). This variability confers an adaptive advantage for some over others, depending on the specifics of the setting. Further, individuals within populations (again, more typically, more complex species) migrate across settings for any number of reasons, including changes in natural conditions such as weather or a lack of food, or as a result of social conflicts, or, particularly in the case of humans, possibly out of a desire for greater novelty or opportunity. New settings may confer an adaptive advantage on some who were less advantaged in the original setting. Further, given a relatively stable environment, these adaptive (or nonadaptive) qualities are then passed across generations through both genetic and epigenetic mechanisms, potentially resulting in morphological, physiological, and behavioral changes in future generations. These phenotypic changes are seen as potential first steps in the evolutionary process because they provide the variability on which natural selection acts. From an evo–devo perspective, although natural selection chooses some species characteristics over others, it is the developmental histories of individuals within a population that determine the characteristics that natural selection then acts on. Ultimately, for species evolution to occur, there must be changes in the genotype, but rather than these genetic changes being seen as the first step in the evolutionary process, from an evo–devo perspective, they are more likely appreciated as a final step in the evolutionary process (West-Eberhard 2005; Lickliter and Schneider 2006; Lamm and Jablonka 2008).

One example of this phylogeny–ontogeny interdependence is seen in the increasing lactose tolerance found in cultures with a history of cattle rearing. The nutritional value of dairy products provided an adaptive advantage, leading to a greater number of surviving offspring and therefore an increasing percentage of those lactose tolerant within cattle-related cultures. In this instance, the evolutionary change was not in the genome per se but rather in the epigenetic mechanisms that regulate those genes affecting the digestion of lactose, in particular, the mechanism that allows lactose tolerance to continue after weaning, the point at which the gene is typically turned off in non-cattle-related cultures. The data on lactose tolerance suggest that, at least for humans, the rate of species evolution may be greater than for other species because of the potential role of the relatively rapid evolution of human cultures. Whether a more rapid evolutionary process ultimately proves advantageous remains to be seen, and it must be emphasized that we are still talking about a process measured in thousands of years but, in any case, such a pattern highlights the interdependence rather than the independence of ontogeny and phylogeny (Newson et al. 2007).

If evo–devo tells developmentalists anything, it is that reductionist explanations of development are at best a misrepresentation of the process of development. They simply do not reflect the bidirectional, probabilistic complexity of the process. The proxy debate is therefore resolved. Systems metamodels are more consistent with our growing understanding of the process than mechanistic metamodels. But evo–devo tells us something else, namely, that ontogeny and phylogeny are interdependent. And it is this interdependence that provides the potential for an integration of DST and EP. It may not be an integration that proponents of either view would fully embrace, but for the rest of us with perhaps somewhat less "ego" involved, there are places to begin to appreciate the compatibilities.

The New (Actually Old) View

Evo–devo is making very clear that reductionist accounts of psychological phenomena, accounts that attempt to differentiate the effects of nature from the effects of nurture and that look for external explanations of developmental patterns, are simply wrong. A new view is needed, one that is consistent with the evo–devo perspective in recognizing the interdependence of ontogeny and phylogeny, that recognizes the interdependence of nature and nurture, that recognizes that development is in large part a self-organizing process, and that this process is best characterized as multidimensional and bidirectional. What is somewhat ironic about the emergence of this new view is that it really isn't that new or, as van Geert (1998) puts it, "We almost had a great future behind us" (p. 143). Van Geert's point is that we have been here before, primarily through the then-emerging views in the United States of developmental theorists such as Werner, Piaget, and Vygotsky in the early 1960s. The works of these developmentalists all focused on trying to establish a true developmental science. Such a developmental science focused on answering the basic questions of what in fact development is all about (as opposed to simply measuring variability or change in behavior), what mechanisms regulate this process, how best can it be studied, and to what degree are both developmental processes and outcomes shared by all members of a species. But van Geert argues that we lost our way:

> The adoption of statistical and design methods led to a certain trivialization of the original questions and helped turn developmental research into a quest for ages and acquisitions and their distributions across specified populations. The reason for this change in course does not lie in the adoption of statistics per se,

but in the adoption of statistics that were designed for different purposes, namely distinguishing populations characterized by some special feature (e.g., a particular treatment or background) and estimating the linear association between the variance of some independent variable on the one hand and some dependent variable on the other. (1998, p. 146)

The (re)emergence of a developmental science consistent with an evo–devo perspective is once again putting us back on track. This developmental science view as summarized by Lewis (2000), Lerner (2006), Cairns et al. (1996), and others, aligns the study of development with the same basic principles of other sciences (e.g., biology, physics, chemistry, and ecology) that also examine self-organizing systems. In particular, a developmental science view argues that development is best understood as a self-organizing, emergent process that, over time, leads to greater organizational complexity that in turn allows the organism a greater degree of adaptiveness. Further, this self-organizing system must not only account for ontogenetic development but must, as well, ensure the reproduction of these self-organizing systems across generations (Stotz 2008).

Despite their sometimes claims to the contrary, I think that there are significant points of integration between Tooby and Cosmides' (Tooby et al. 2003, 2005; Tooby and Cosmides 2005) evolutionary psychological perspective and Gottlieb's (1997, 2002, 2003, 2007) developmental systems perspective, largely because both are systems based and both address the fundamental developmental issues previously listed. And it is the integration of these two perspectives that can serve as the foundation for a complete developmental science, one that integrates phylogeny with ontogeny and offers a more integrated view of the developmental process.

Although EP and DST are often portrayed at odds with each other by their respective advocates, they really are better viewed as complementary rather than antagonistic. The areas that each focuses on tend to be the areas that the other gives less attention to, and, as such, putting the two together provides a much more complete and compelling picture of both ontogeny and phylogeny than either provides alone. In fact, as Partridge and Greenberg (2010) note in their discussion of DST, it and EP are the only orientations capable of such a grand synthesis:

Perhaps it is because comparative psychologists, developmental psychobiologsts, contemporary developmental psychology, and evolutionary psychology all have as a central focus the integration of phenomena across multiple levels of analysis (i.e., biology, behavior, local ecology, culture) that there is a recognition of the need for a broad synthesizing theory. (p. 178)

Probably the best way to demonstrate this complementariness is to consider three common points that each perspective addresses. By doing so, I hope to show that in spite of sometimes over-the-top characterizations – EP describing DST as adhering to the "standard social science model" and DST characterizing evolutionary psychologists as "genetic preformationists" – what each, in fact, is advocating isn't nearly as different from the other as each claims it to be. The three points are:

1. the relationship of phylogeny and ontogeny, including the mechanisms regulating each;
2. the nature of our species specific characteristics, in particular, the degree to which our cognition is modularized; and,
3. the degree to which cognitive systems are open or closed, that is, the question of plasticity.

Let's take each in turn, first ontogeny and then phylogeny. As far as the modern synthesis was concerned, there was no relationship between ontogeny and phylogeny because evolution was a function of the genome and the genome was not influenced by an individual's developmental history. Rather, the focus was on mutation and population genetic drift and recombination, presumably all random processes. But both EP and DST place much more emphasis on adaptation as the primary mechanism influencing natural selection, and, as such, both recognize that one's developmental history, that is, one's ontogeny, does hold the potential to influence the evolution of that organism's species.

Both DST and EP argue that variability between organisms results in some expressing more adaptive behaviors than others. Further, this adaptiveness is seen as a goodness of fit between the organism's developmental status (i.e., the cumulative developmental history of the organism's genome in a synergistic bidirectional interaction with that organism's environment) and what EP refers to as the organism's *environment of evolutionary adaptiveness.* DST makes much the same argument. In both cases, natural selection acts on adaptive or nonadaptive phenotypes to increase the likelihood of the first across generations and decrease the likelihood of the second. But whereas EP tends to focus on the last step in the process (i.e., natural selection leading to changes in the genome), DST describes a multistep process of which changes in the genome is the final step. The two do not contradict each other so much as they complement each other. The emphasis on change in the genome highlights the fact that permanent change must ultimately involve changes in the genome, and the emphasis on the precursors of genetic shift emphasizes the fact that the evolutionary process is continuous and is active in selection even before change in the genome itself occurs. EP doesn't argue against a multistep evolutionary process so much as it simply focuses on the last step. Nevertheless, EP does acknowledge the role of these "additional systems of inheritance." In particular, Tooby et al. (2003) note that:

If it must often be the case that there is a correlation of conditions among adjacent generations in certain respects (e.g., if a mother faces an exceptionally competitive, predatory, or food limited environment) then there is an increased probability that offspring will too – and with some decay function, that subsequent generations will as well. This would select for additional systems of inheritance that could transmit regulatory signals from immediately preceding generations. The function of these signals is to help send individual development along pathways better suited to the conditions it is likely to face in life. Given the operation of such systems, phenotypes would be partly inherited (i.e., cross-generationally regulated with heightened parent-offspring similarity) in a way not attributable to DNA-sequence differences. (p. 859)

There is little difference between this statement by Tooby et al., and Gottlieb's (1997) description of events in the second step of his systems based evolutionary perspective:

Thus, behavioral and morphological phenotypic changes can be immediately instigated by a change in an individual's developmental conditions. In our view, a change in developmental conditions activates heretofore quiescent genes, thus changing the usual developmental process and resulting in an altered behavioral or morphological phenotype. Consequently, stage II in the evolutionary pathway holds that the new environmental relationship brings out latent possibilities for morphological-physiological change in advance of the usual criterion of evolution: a change in structural genes or gene frequencies in the population. (p. 159)

So, both EP and DST acknowledge the role of nongenetic mechanisms of inheritance, that natural selection acts on the degree to which organisms are successful in adapting to changing circumstances, and that ultimately sustaining evolutionary change must involve change in the genome. Such a view defines an essential role for ontogeny in species evolution. Such a view also offers an interesting reinterpretation of EP's argument that contemporary humans are using a stone-age mind in trying to deal with twenty-first-century realities. We may well be using a stone-age genome, but that doesn't mean that there have not been and continue to be significant epigenetic morphological, physiological, and behavioral changes in our species as a result of our efforts to adapt to our cumulative species history; changes that at some point may or may not eventually lead to changes in our stone-age genome. Maybe we need to give a closer look to what EP refers to as the by-products of species evolution; they might just play a larger role in the course of human history than EP suggests. In fact, Pinker himself in his most recent book (2011) makes the same argument when he notes that there has been a significant decrease in the level of human hostility over recorded history, a time period much too short to suggest that such a change reflects actual changes in the human genome.

Now let's consider the second potential point of contact between EP and DST, the issue of species characteristics. DST offers a model through which the process of development occurs. As mentioned several times already, it is a process that is characterized as multilevel, bidirectional, and self-organizing. The outcome of this self-organizing process reflects the emergent synergistic products of the interdependent actions of nature and nurture. Neither nature nor nurture has a privileged position in the process; both influence the likelihood, form, and function of any particular product. Each can serve to influence the likelihood of either stability over time or change over time. From this perspective, the fact that we usually end up with two eyes, one nose, ten fingers, ten toes, etc. reflects both the stability of the genotype for these characteristics and, equally, the stability of the prenatal environment. In other words, both the nature and the nurture of these characteristics appear highly canalized. However, because the focus of DST has been on modeling the process through which development is believed to occur, it has devoted considerably less time to detailing what the products of this process might actually look like.

On the other hand, EP does devote considerable time to detailing the products of the evolutionary process when it argues for the presence of "hundreds or perhaps even thousands" of cognitive modules that constitute the repertoire of phylogenetic adaptations that have proved successful over our species history. These modules are believed ultimately coded in the genome, serve to regulate to some degree and in some manner the architecture of the brain within an individual, and are expressed at different points in the life span. So some modules, such as those influencing attachment, are expressed early in life because they serve an immediate adaptive purpose whereas others, perhaps those dealing with the establishment of sexual intimacy among peers, might not be expressed until adolescence or the adult years. In some cases the modules are described as "innate" and in other cases as "instincts." But in spite of the highly detailed description of a number of these modules, particularly those dealing with gender differences and sexual behavior, EP has really not offered a mechanism through which these modules come into being or become coded in the genome or then eventually come to be expressed through behavior.

So, to overstate the matter a bit, DST offers a process without a description of products and EP offers products without a description of process. How then can these two perspectives be seen as complementing each other? The answer to this question requires first a recognition that in fact both DST and EP view the interplay of ontogeny and phylogeny from an evo–devo perspective, and second, that we need to reconsider what it means to say that something is innate. Let's take each in turn.

The link between evo–devo and DST is easy to see. They are essentially one and the same. Both emphasize the role of adaptation in natural selection; both recognize phylogeny as the cumulative product of successive ontogenies; both recognize genetic and nongenetic (i.e., epigenetic) cross-generational mechanisms of inheritance; and both argue that the influence of nature and nurture is interdependent and synergistic. It makes no sense to talk about one as independent of the other. But, as previously noted, this is pretty much what EP says as well. EP is an adaptationist model; all of its proposed modules represent successful adaptations to selective pressures. Because natural selection acts on or selects certain adaptations, it is acting on ontogenetic efforts to maintain a degree of equilibrium within a particular context. EP does in fact recognize epigenetic as well as genetic mechanisms of inheritance, although it does place more emphasis on the genetic mechanisms. And, at least with respect to the acquisition of modules, EP argues that the influences of nature and nurture are also interdependent. Said another way, the differences between DST and EP with respect to evolutionary mechanisms are better appreciated as ones of emphasis rather than ones of type, and, as such, there is room here for further research on better defining the links between the two models.

The larger problem comes in when we now consider how these evolutionary products are expressed. DST is sometimes portrayed, unfairly, as arguing that we are infinitely pliable, totally open systems. EP, on the other hand, is sometimes portrayed, equally unfairly, as arguing that individual development is no more than the simple, predetermined, unfolding of a bunch of modularized adaptations that seem to have made life easier thousands and thousands of years ago. Neither characterization is truly accurate. Both recognize that we enter the world as no more lumps of clay than highly scripted computer programs. So what do we come into the world with and what then regulates individual development? The answer to this question rests first on what it means for something to be innate.

To refer to something as innate simply means that it is present at birth. The meaning of the word technically carries no more baggage although, because it is often used interchangeably with the term instinct, it sometimes is seen as carrying the same baggage as the term instinct. Instinct, on the other hand, conveys the image of moths driven to the flame, a predetermined course of action unaffected by context except for the necessity of the flame to release the instinct. Instinct is a term more related to the classic debate, and, as such, it should enjoy the same fate as the classic debate. Innate, on the other hand, could serve as the meeting place for DST and EP.

We don't enter the world empty handed. We already carry with us data carried in the genome as they are interpreted during the prenatal process. EP has typically placed more emphasis on the influence of these data during this prenatal interplay and DST on the interpreters (that is, self-organizing prenatal context) of the data. But in fact you can't have one without the other, and each acquires meaning only in the context of the other. So why should we be so afraid when it is claimed (Baillargeon et al. 1985; Baillargeon 2008) that object permanence may be present at birth or that newborns can add (Spelke 1998; Spelke and Kinzler 2009)? We are afraid because we incorrectly have attached the baggage of instinct onto innateness and therefore incorrectly assume that something present at birth is a done deal. But it isn't. Very young infants may be able to recognize discrepancy in the number of small objects now present in their visual fields from what was presented previously but they clearly aren't yet able to successfully add a list of 10 six-digit numbers, and they won't be able to do so for some time, and then only if something in the way of instruction is offered to them. Similarly, a young infant's seeming surprise when the train doesn't come out of the tunnel as expected suggests that newborns enter the world with some cognitive mechanisms allowing the expression of expectancy (as well as violations of the same), but this same infant is not likely going to be pulling the blanket off of the hidden object for some months to come. Even setting aside the debate as to whether, for example, newborn addition and newborn object permanence are even in some way reflective of the same processes in more mature children and adults, the fact remain that there is clear evidence of some number of rudimentary abilities being present at birth, and both DST and EP support such a position. How could it be otherwise?

The real question then is the regulation, both prenatally and postnatally, of whatever number of and in whatever form these modules might exist. Gottlieb's work on the prenatal determinants of ducklings following their mothers soon after birth makes clear the need to broaden our conception of antecedents, especially in terms of recognizing nonobvious ones. And the evolutionary record makes clear that each generation does not start from scratch; there are species-typical characteristics that we share, and these characteristics do reflect our evolutionary history. But EP is typically silent on the mechanisms regulating the expression of these species-typical characteristics, leaving the impression, particularly among its critics, that the EP simply sees the regulatory mechanism as genetic and predetermined. But to be consistent with its arguments as to how modularized adaptations are initially acquired, EP needs to recognize that the mechanisms regulating expression need to be consistent with those regulating acquisition, and when this happens, we do get pretty close to the multilevel, bidirectional mechanisms championed by DST. Some may be innate in the sense that they are expressed at birth or very soon thereafter; others may not be expressed until later, but in either case, the expression must depend on a synergistic interdependence between nature and nurture, and this interdependence may not be obvious. So, if in fact a fear of snakes is virtually universal, even among people who have never had contact with them, then the issue becomes one of identifying the nonobvious as well as the obvious contextual antecedents that interact with the genome rather than simply claiming that ophidiophobia is genetically predetermined. Or when we find seemingly stereotypical sex role behavior across what appear to be widely divergent cultural settings, then we reconsider what measures we are using to define cultures as similar or distinct and we acknowledge that just as the universality of gravity has influenced how we move as a species, so too might the realities of procreation influence, to one degree or another, how we behave as males and females. EP and DST complement each other when we look at the middle rather than the extremes of each position, or, even worse, the caricatures of each sometimes offered by the other.

Just as the first two issues are not independent of each other, neither is the third, the question of plasticity. How much of our fate is already set through our evolutionary heritage and how much of it is really to be etched onto a blank slate? In one sense, this is the classic debate nature–nurture question, but when considered from the perspective of the new debate, the question may have a different answer. Here too, critiques of ER and DST exaggerate the respective positions. In fact, EP does not say that our lives are little more than the playing out of some sort of predetermined Greek tragedy and DST does not say that all it takes to be anyone and do anything is essentially to wish upon a star.

For Tooby and Cosmides (2005), the highly modularized structure of the brain leads not to inflexibility but to just the opposite because individual modules, because of their hypothesized independence, are able to coordinate with each other in novel ways as the adaptational demands of a situation require. In fact, for EP, if there wasn't this flexibility built into the brain's architecture, then organisms could never demonstrate the novel adaptive behaviors on which natural selection acts. Without flexibility, there can be no evolution. As they put it, "Armed with this insight, we can lay to rest the myth that the more evolved organization the human mind has, the more inflexible its response. Interpreting the emotional expressions of others, seeing beauty, learning language, loving your child – all these enhancements to human mental life are made possible by specialized neural programs built by natural selection" (p. 17).

And Gottlieb (1997) makes virtually the same argument when he talks about the development of *behavioral neophenotypes*, that is, adaptational epigenetic changes that have not yet been incorporated into the organism's genome. In particular, he argues that:

> The creation of behavioral neophenotypes is necessarily dependent on the existence of some degree of behavioral plasticity or adaptability. Thus, the determinants of behavioral plasticity are an important consideration. One key limiting component of plasticity is the nervous system, particularly the brain, and the other is the developing organism's early experiences. These two components are in lockstep: Larger brained species can make more of their early experience, and early experiences affect the maturation and size of the brain. Thus, the most conspicuous developmental route to increasing behavioral plasticity and creating behavioral neophenotypes is through early experiential alterations (including nutrition) that have positive effects on enhancing the maturation of the brain. (pp. 151–2)

So, contrary to claims of critics of DST and EP, critiques that again depict each as more extreme than in fact it is, there does seem to be a meeting ground with respect to plasticity. Both perspectives acknowledge the essential role that plasticity plays in allowing for the novel adaptive behavior that natural selection acts on, and both support the hypothesis that the more highly cognitively evolved the species, the greater the potential for plasticity and therefore the greater potential for successful adaptations (i.e., survival). DST does go a step further in arguing that the immature organisms may be more plastic than mature ones, a hypothesis that EP doesn't dispute so much as it doesn't consider (with the exception of Bjorklund's (2006) work).

A second possible meeting place with respect to plasticity concerns the organization of elements in a system. EP argues that our ability to adapt presumes effective strategies for dealing with both expected and unexpected contingencies and the only way this is possible is by having a highly organized set of domain-specific modules. The organization of these modules allows for the quick recognition of the situation and the quick application of the correct modules to the situation:

For example, what counts as a "good" mate has little in common with a "good" lunch or a "good" brother. Designing a computational program to choose foods based on their kindness or to choose friends based on their flavor and the aggregate calories to be gained from consuming their flesh suggests the kind of functional incompatibility issues that naturally sort human activities into incommensurate motivational domains. Because what counts as the wrong thing to do differs from one class of problems to the next, there must be as many domain-specific subsystems as there are domains in which the definitions of successful behavioral outcomes are incommensurate. (Tooby and Cosmides 2005, p. 48)

An organization that involves a motivation subsystem and a valuation subsystem as well as domain-specific content modules is, for EP, the necessary cognitive structure that offers an adequate and comprehensive explanation of how we have managed to survive as a species for as long as we have. EP is not very specific as to how the system organizes itself or for that matter what the actual anatomical structure of the brain would need to be to account for such organization. Given the fact that EP envisions these modular systems existing in the particular forms that neuronal networks create during both prenatal and postnatal development rather than residing in any particular place in the brain, one important research question would be to determine how that data present in the genome, in bidirectional interaction with various levels of environments then does manage to structure the neuronal networks that will serve to make possible appropriate adaptive responses, that is, to make sure that one doesn't confuse what makes a good meal from what makes a good sibling.

If the presumed multileveling structure of modules has a bit of a familiar ring to it, you're right. Isn't this question of organization also the focus of DST in particular and systems theory in general? In particular, for both DST and EP, as well as for systems theories in general (Partridge and Greenberg 2010) the implication is that the organization of the elements of a system is an emergent property of that system itself and does not require an independent antecedent. Further, for both, this self-organizing process creates a tiered system that actually allows for greater flexibility that in turn offers the possibility of more effective adaptability, a key characteristic of both EP and DST. Partridge and Greenberg (2010) characterize such self-organizing systems as displaying "molar level stability and micro level instability," which allows these systems to be "quite adaptable to changing environmental pressures and contingencies making them ideal for flourishing under principles of natural selection" (p. 172).

I would be the first to admit that my effort to document the ways in which EP and DST complement each other is far from perfect. But if we are to truly understand the developmental process, then we need to begin to view the process not from competing perspectives but rather from complementary perspectives. Whether there really are hundreds or even thousands of evolutionary-based "modules" is, ultimately, really not the issue; rather, the issue is recognizing that the cumulative evolutionary history of a species has an impact on the ontogeny of any particular generation of that species. And, at the same time, it is also to recognize that this evolutionary influence is far from defined independently of the particulars of that generation's ontogeny. If 5-year olds can get beyond dichotomies, surely we can as well.

Studying Everything at Once

How then is it possible to study everything at once? If both ontogeny and phylogeny involve multicomponent, multilevel, bidirectional mechanisms, how is it possible to even talk about what is causing what? There just aren't enough introductory psychology students in the world to pursue the study of everything at once. And how can you possibly prepare developmentalists to study everything at once? We would need to spend the rest of our lives in graduate school to pick up the necessary skills in all of the relevant disciplines. Fortunately things may not be quite so dire. Two possible directions for study do present themselves. The first concerns some agreement on what we study, and the second concerns how we actually go about doing it.

A synthesis of the complementary perspectives in the "new debate" must lead us to the conclusion that what we study as developmentalists is a developing system, and, like all systems, the meaning of the elements of that system can be understood only when functioning interdependently with the other elements, at all levels, of the system. Like all systems, development viewed from a systems perspective is best understood as a self-regulating process that, as is true of all other organic systems, increases in complexity over time. This is not to suggest that antecedents, that is, material and efficient causes, do not influence the system but rather to suggest that the nature of such influence is significantly determined by the formal properties of the system itself. In other words, in a nonlinear system, there is no one-to-one correspondence between "cause" and "effect." Seemingly small antecedents, introduced at a particular time, might have a large effect but the same antecedent, introduced at a different time, might have a negligible effect or even an opposite effect, depending on the state of the system at that moment. For example, with respect to genetics, the essential issue is not simply the presence or absence of any particular gene (or module for that matter) but rather its expression, and this expression is a reflection of the system of which that genome is a component (Meaney 2001; Stotz 2008; Meaney 2010). As Overton (2010) puts it, "The organism is, thus, an *embodied cognitive-cognate-emotional system* – not driven by a split-off brain nor by a split-off culture – that develops through bidirectional reciprocal nonlinear interactions with its biological, physical, and social worlds" [italics in original] (p. 3).

One of these formal properties is the system's efforts to maintain a degree of dynamic equilibrium both among the elements of the system and between the system and its milieu. This equilibrium is seen as dynamic because, unlike our homeostatic mechanisms that regulate such functions as body temperature, dynamic efforts periodically result in shifts in the formal organization of the system. These stage or phase shifts are seen as providing the developing system a greater degree of adaptability and therefore a higher probability of maintaining a degree of equilibrium, that is, the reorganization of the system leads to the emergence of stable, novel, measurable, new modes of functioning, as is the case when children go from crawling to walking or with the emergence of language as a communication and symbolic tool or with the emergence of various forms of social organization over a species' evolutionary history (Greenberg et al. 2006). One major research question about such a dynamic process is the relative valence of the elements of the system at any particular point in the process and whether these valences change as new properties of the system emerge (Magnusson and Cairns 1996). From such a developmental systems perspective, it is possible to recognize that the process of ontogeny and the process of phylogeny appear to follow the same set of structural and functional dynamics.

This image of development as a nonlinear, emergent process has always been viewed as highly conceptual and therefore, to some, of little empirical value. This was certainly the case in Brainerd's critique (1978) of Piaget's theory, in particular the claim of a lack of empirical referents for predicting and measuring stage transitions. But the growing recognition of the value of general nonlinear dynamic systems theory may be providing an answer to the types of arguments made by Brainerd. Such a model is not specific to development but more generally to any system that displays *punctuated equilibrium*, that is, periods of stability interrupted by relatively abrupt shifts in organization. The model has been used to offer insights into evolutionary change, changes occurring during prenatal development, and, more generally, processes within the disciplines of both chemistry and the life sciences (van der Maas and Hopkins 1998; Partridge and Greenberg 2010).

Consider, for example, an early study looking at Piagetian stage transitions related to the conservation. Van der Maas and Molenaar (1992) were able to identify a four-step transition sequence. Those not yet capable of understanding the task showed some form of guessing in their responses. A second group seemed to understand the task but were misled by perceptual cues. A third group gave conserving responses but could offer little in the way of a logical rationale, whereas the fourth group both consistently gave a conserving response as well as appropriate logical explanations. From a dynamic systems perspective, the sequence can be characterized as involving a perceptual and a cognitive factor. The first group demonstrated neither, the last group both, and the two middle groups in-between values. It is the third group that is seen as best representing the transition or "cusp" phase in that they faced two conflicting strategies or argumentations which brings them "far from equilibrium." More generally, during stage or phase transitions, we would expect to see a greater degree of response variability, longer latencies of response as the child attempts to reconcile the conflicting perceptual and cognitive cues, and as the transition is nearing resolution, a sudden jump in response certainty:

> Suffice it to say, that for our purposes, applied to psychology, we can identify emergent phenomena by several criteria: they display radical novelty – features not present in the underlying complex system; they display coherence or correlation – they have a unit over time; they exist at the global or macro level and not at all at the underlying micro level; they are dynamical, arising as a result of the dynamic interplay of the underlying micro events; and they are ostensive – they really exist and are observable. (Partridge and Greenberg 2010, p. 174)

One essential consideration in the empirical verification of such stage transitions is that the data must be collected longitudinally and at frequent-enough intervals to capture the steps in the transition process. In other words, the primary focus of developmental research should be on intra-individual rather than on interindividual variability because the former actually measures change whereas the latter only compares current status (Molenaar 2007). Although the traditional focus on interindividual variability is typically justified in terms of generalizing to a population, van Geert (1998) argues that such generalizations are as easily done when the focus is on intra-individual variability. In particular, he argues that the best way to understand a presumably universal dynamic process is to first sample the possible states of the process across a population in order to determine the possible "state spaces" of the process across a range of initial conditions. Once these state spaces are identified, then the next task becomes following these intra-individual trajectories frequently enough to map the dynamics of the process. The final step involves " sampling a representative range of trajectories, representative in the sense that they cover the set of possible or likely trajectories, not in the sense of proportional representation as in population sampling" (van Geert, 1998, p. 155). This greater focus on intra-individual development again puts "time" back into the developmental equation, not as a causal mechanism but rather as a ruler against which dynamic processes can be observed and measured and, consistent with van Geert's argument that developmental research over the past few decades has been anything but developmental, the recognition of the insights provided by the grand developmental theorists such as Piaget, Vygotsky, and Werner.

In the broadest sense, such a developmental systems approach requires what Overton (2006) refers to as a relational metatheory and metamethod. This relational concept is, according to Overton, best represented in Escher's famous picture of two hands, each simultaneously drawing the other. One cannot exist without the other and so too in relational metatheory and metamethod, individual elements exist not as independent atomistic elements but as related elements in a system:

> Matter and society represent systems that stand in an identity of opposites. To say that an object is a social object in no way denies that it is matter; to say that an object is matter in no way denies that it is social. The object can be analyzed from either a social or a physical standpoint, and the question for synthesis becomes the question of what system will coordinate these two systems. Arguably, the answer is that it is *life* [italics in original] or living systems that coordinate matter and society. (Overton 2006, p. 37)

As such, the synthesis of the biological and the social is the *psychological organism* (Goldhaber 1986), and, by extension, the synthesis of any two of these perspectives becomes the third. Any of the three can be the focus of study, but the findings from such research will be meaningful only when considered within the context of the other two. Or, as Kagan (2006) wryly put it, neuroscientists can put as many electrodes on his head as they like; they still won't be able to successfully predict whether he is going to decide to have soup or salad for lunch.

So, when all is said and done, here is what the "proper" study of development has to offer:

- A nonlinear systems view of the process of development that recognizes and draws on theoretical and methodological parallels with other disciplines valuing a dynamic systems perspective.
- An appreciation for the value of finding the complementariness between perspectives rather than their antagonisms.
- An approach to methodology that emphasizes intra-individual change rather than interindividual status.

- An understanding that at least equal scholarly effort be devoted to the study of developmental patterns (i.e., formal and final causes) as to the study of developmental antecedents (i.e., efficient and material causes), especially because the two are interdependent.
- An appreciation of the fact that observation and interpretation (i.e., method and theory) are not independent but rather complementary and interdependent.
- A recognition of the fact that ontogeny and phylogeny are both significant influences on individual development and are interdependent.
- And, a recognition, finally, once and for all, that atomistic, reductionist approaches to the study of development are simply incorrect, given what we now understand about the process of development; inevitably they do more harm than good.

Truth be told, none of this is really new; Wohlwill (1973) pretty much said the same thing years ago, but it doesn't hurt to be reminded of things once in a while, and hopefully this book has served that purpose.

It really is incredible when you think about it. There may actually be a way to close the gap after all.

Bibliography

Alford, J. R., C. L. Funk, and J. R. Hibbing (2005). "Are political orientations genetically transmitted." *American Political Science Review* 99(2): 153–67.

Alphonse, L. (2011). "Note to parents: Stop trying so hard." *Shine on Yahoo.* Available at (http://fe1.shine.lifestyles.fy11.b.yahoo.com/event/mothersday/note-to-parents-stop-trying-so-hard-2479952). (Accessed May 2).

Anandalakshmy, S. and R. E. Grinder (1970). "Conceptual emphasis in the history of developmental psychology: Evolutionary theory, teleology, and the nature-nurture issue." *Child Development* 41: 1113–23.

Anastasi, A. (1958). "Environment, heredity and the question 'how?'" *Psychological Review* 65(4): 197–208.

Anastasi, A. and J. P. Foley (1948). "A proposed reorientation in the heredity-environment controversy." *Psychological Review* 55(3): 239–49.

Arthur, W. (2002). "The emerging conceptual framework of evolutionary developmental biology." *Nature(London)* 415(6873): 757–64.

Baer, D. M. (1970). "An age-irrelevant concept of development." *Merrill-Palmer Quarterly* 16: 238–45.

Baillargeon, R. (2002). "The acquisition of physical in infancy: A summary of eight lessons." In *Blackwell's Handbook of Childhood Cognitive Development.* Oxford, UK: Blackwell, pp. 47–83.

Baillargeon, R. (2008). "Innate ideas revisited: For a principle of persistence in infants' physical reasoning." *Perspectives on Psychological Science* 3(1): 2.

Baillargeon, R., E. S. Spelke, and S. Wasserman (1985). "Object permanence in five-month-old infants." *Cognition* 20: 191–208.

Barnett, W. S. (2008). *Preschool Education and Its Lasting Effects: Research and Policy Implications.* Rutgers, NJ: Rutgers University Press.

Baumrind, D. (1993). "The average expectable environment is not good enough: A response to Scarr." *Child Development* 64(5): 1299–1317.

Bell, R. (1968). "A reinterpretation of the direction of effects in studies of socialization." *Psychological Review* 75(2): 81–95.

Belsky, J. and M. Pluess (2009). "The nature (and nurture?) of plasticity in early human development." *Perspectives on Psychological Science* 4(4): 345–51.

Bjorklund, D. (2003). "Evolutionary psychology from a developmental systems perspective: Comment on Lickliter and Honeycutt (2003)." *Psychological Bulletin* 129(6): 836–41.

Bjorklund, D. F. (2006). "Mother knows best: Epigenetic inheritance, maternal effects, and the evolution of human intelligence." *Developmental Review* 26: 213–42.

Bjorklund, D. F. (2007). *Why Youth Is Not Wasted on the Young.* Malden, MA: Blackwell.

Bjorklund, D. F. and K. K. Harnishfeger (1990a). "Children's strategies: A brief history." In *Children's Strategies: Contemporary Views of Cognitive Development*, edited by D. F. Bjorklund. Hillsdale, NJ: Lawrence Erlbaum, pp. 1–16.

Bjorklund, D. F. and K. K. Harnishfeger (1990b). "Children's strategies: Their definition and origins." In *Children's Strategies: Contemporary Views of Cognitive Development*, edited by D. F. Bjorklund. Hillsdale, NJ: Lawrence Erlbaum, pp. 309–24.

Bjorklund, D. F. and A. D. Pellegrini (2000). "Child development and evolutionary psychology." *Child Development* 71(6): 1687–1708.

Bjorklund, D. F. and A. D. Pellegrini (2002). *The Origins of Human Nature: Evolutionary Developmental Psychology.* Washington, DC: American Psychological Association.

Blasi, C. H. and D. F. Bjorklund (2003). "Evolutionary developmental psychology: A new tool for better understanding human ontogeny." *Human Development* 46: 259–81.

Blumberg, M. S. (2009). *Freaks of Nature.* Oxford, UK: Oxford University Press.

Bouchard, T. J. J. (2009). "Strong inference." In *Experience and Development*, edited by K. McCartney and R. A. Weinberg. New York: Psychology Press, pp. 39–60.

Brainerd, C. J. (1978). "The stage question in cognitive-developmental theory." *Behavioral and Brain Sciences* 2: 173–213.

Bronfenbrenner, U. (1979). *The Ecology of Human Development.* Cambridge, MA: Harvard University Press.

Bronfenbrenner, U. (1999). "Environments in developmental perspective: Theoretical and operational models." In *Measuring Environments Across the Life Span: Emerging Methods and Concepts*, edited by S. Friedman and T. D. Wachs. Washington, DC: American Psychological Association, pp. 3–28.

Bronfenbrenner, U. (2001). "The bioecological theory of human development." In *International Encyclopedia of the Social and Behavioral Sciences*, edited by N. J. Smelser and P. B. Baltes. New York: Elsevier, Vol. 10, pp. 6963–70.

Bronfenbrenner, U. and S. Ceci (1994). "Nature-nurture reconceptualized in developmental perspective: A bioecological model." *Psychological Review* 101(4): 568–86.

Bronfenbrenner, U. and G. Evans (2000). "Developmental science in the 21st century: Emerging questions, theoretical models, research designs and empirical findings." *Social Development* 9(1): 115–25.

Browne, K. R. (2004). "Women in the workplace: Evolutionary perspectives and public policy." In *Evolutionary Psychology, Public policy and Personal Decisions*, edtied by C. Crawford and C. Salmon. Mahwah, NJ: Lawrence Erlbaum, pp. 275–92.

Burian, R. M. (2005). *The Epistemology of Development, Evolution, and Genetics.* Cambridge: Cambridge University Press.

Buss, D. (2001). "Human nature and culture: An evolutionary psychological perspective." *Journal of Personality* 69(6): 955–78.

Buss, D. M. (2009). "The great struggles of life." *American Psychologist* 64(2): 140–8.

Butcher, L., J. Kennedy, and R. Plomin (2006). "Generalist genes and cognitive neuroscience." *Current Opinion in Neurobiology* 16(2): 145–51.

Cairns, R. B., G. H. Elder, and E. J. Costello, Eds. (1996). *Developmental Science.* Cambridge: Cambridge University Press.

Calman, L. J. and L. Tarr-Whelan (2005). *Early Childhood Education for All: A Wise Investment.* New York: Legal Momentum. Available at (http://www.legalmomentum.org/fi/pdf/FamilyInitiativeReport.pdf).

Carmichael, L. (1925). "Heredity and environment: Are they antithetical?" *Journal of Abnormal Psychology* 20: 245–60.

Carmichael, L. (1926). "What is empirical psychology?" *American Journal of Psychology* 37(4): 521–527.

Caspi, A., J. McClay, T. E. Moffitt, J. Mill, J. Martin, I. W. Craig, A. Taylor, and R. Poulton (2002). "Role of genotype in the cycle of violence in maltreated children." *Science* 297(5582): 851–4.

Caspi, A., et al. (2003). "Influence of life stress on depression: moderation by a polymorphism in the 5-HTT gene." *Science* 301(5631): 386–9.

Causey, K., A. Gardiner, and D. F. Bjorklund (2008). "Evolutionary developmental psychology and the role of plasticity in ontogeny and phylogeny." *Psychological Inquiry* 19(1): 27–30.

Chiszar, D. A. and E. S. Gollin (1990). "Additivity, interaction, and developmental good sense." *Behavioral and Brain Sciences* 13: 124–5.

Chua, A. (2011). *Battle Hymn of the Tiger Mother.* New York: Penguin Press.

Confer, J. C., J. A. Easton, D. S. Fleischman, C.D. Goetz, D. M. G. Lewis, C. Perilloux, and D. M. Buss (2010). "Evolutionary psychology: Controversies, questions, prospects, and limitations." *American Psychologist* 65(2): 110–26.

Cooper, R. S. (2005). "Race and IQ." *American Psychologist* 60(1): 71–6.

Cosmides, L. and J. Tooby (2001). "Unraveling the enigma of human intelligence: Evolutionary psychology and the multimodular mind." In *The Evolution of Intelligence,* edited by R. J. Sternberg and J. C. Kaufman. Hillsdale, NJ: Lawrence Erlbaum, pp. 145–98.

Cosmides, L. and J. Tooby (n/d/). *Evolutionary Psychology: A Primer, Center for Evolutionary Psychology,* University of California, Santa Barbara.

Cronbach, L. J. (1957). "The two disciplines of scientific psychology." *American Psychologist* 12: 671–684.

Cronbach, L. J. (1969). "Heredity, environment, and educational policy." *Harvard Educational Review* 39(2): 338–47.

Dagg, A. (2004). *"Love of Shopping" is Not a Gene: Problems With Darwinian Psychology.* Montreal, Canada: Black Rose Books.

Davies, A. P. C. and T. K. Shackelford (2008). Two human natures: How men and women evolved different psychologies. In *Foundations of Evolutionary Psychology,* edited by C. Crawford and D. Krebs. New York: Lawrence Erlbaum, pp. 261–82.

Detterman, D. (1990). "Don't kill the ANOVA messenger for bearing bad interaction news." *Behavioral and Brain Sciences* 13: 131–2.

Duntley, J. D. and D. M. Buss (2012) "The evolution of stalking." *Sex Roles* 66(5–6): 311–27.

Ehrlich, P. and M. Feldman (2003). "Genes and culture." *Current Anthropology* 44(1): 87–107.

Elder, G. H., Jr. (1974). *Children of the Great Depression.* Chicago, IL: University of Chicago Press.

Elder, G. H., Jr. (1986). "Military times and turning points in men's lives." *Developmental Psychology* 22: 233–245.

Elder, G. H., Jr. (1995). "The life course paradigm: Social change and individual development." In *Examining Lives in Context,* edited by P. Moen, G. H. J. Elder and K. Luscher. Washington, DC: American Psychological Association, pp. 101–39.

Elder, G. H., Jr. (1998). "The life course as developmental theory." *Child Development* 69(1): 1–13.

Elder, G. H., Jr. and J. Z. Giele (2009). "Life course studies: An evolving field." In *The Craft of Life Course Research,* edited by G. H. J. Elder and J. Z. Giele. New York: Guilford, pp. 1–24.

Elder, G. H., Jr. and T. K. Hareven (1993). "Rising above life's disadvantage: From the great depression to war." In *Children in Time and Place: Developmental and Historical Insights,* edited by G. H. Elder Jr., J. Modell, and R. D. Parke. Cambridge: Cambridge University Press, pp. 47–72.

Elder, G. H., Jr., V. King, and R. D. Conger (1996). "Intergenerational continuity and change in rural lives: Historical and developmental insights." *International Journal of Behavioral Development* 19(2): 433–55.

Elkind, D. (1969). "Piagetian and psychometric conceptions of intelligence." *Harvard Educational Review* 39(2): 319–37.

Elkind, D. (1981). *The Hurried Child: Growing Up Too Fast Too Soon.* Reading, MA: Addison-Wesley.

Elkind, D. (1987). *The Miseducation of Children: Superkids at Risk.* New York: Knopf.

Evans, G. (2004). "The environment of childhood poverty." *American Psychologist* 59(2): 77–92.

Evans, G. and K. English (2002). "The environment of poverty: Multiple stressor exposure, psychophysiological stress, and socioemotional adjustment." *Child Development* 73(4): 1238–48.

Falk, R. (2000). "The gene – A concept in tension." In *The Concept of the Gene in Development and Evolution,* edited by P. J. Beurton, R. Falk, and H.J. Rheinberger. Cambridge: Cambridge University Press, pp. 317–48.

Fancher, R. E. (2009). "Scientific cousins: The relationship between Charles Darwin and Francis Galton." *American Psychologist* (Special Issue: *Charles Darwin and Psychology, 1809–2009*) 64(2): 84–92.

Fisher, R. A. (1914). "Some hopes of a eugenist." *Eugenics Review* 5: 309–315.

Fisher, R. A. (1918). "The correlation between relatives on the supposition of Mendelian inheritance." *Transactions of the Royal Society of Edinburgh* 52: 399–433.

Galton, F. (1869). *Heredity Genius.* London: Macmillan.

Geary, D. C. (2006). "Evolutionary developmental psychology: Current status and future directions." *Developmental Review* 26(2): 113–19.

Geary, D. C. and D. F. Bjorklund (2000). "Evolutionary developmental psychology." *Child Development* 71: 57–65.

Gesell, A. and F. L. Ilg (1949). *Child Development.* New York: Harper & Row.

Gianoutsos, J. (2006). "Locke and Rousseau: Early childhood education." *The Pulse* 4(1): 1–23.

Gillette, A. (2007). *Eugenics and the Nature-Nurture Debate in the Twentieth Century.* New York: Palgrave Macmillian.

Gladwell, M. (1998). Do parents matter? *The New Yorker,* Aug. 17, pp. 54–5.

Goldberg, A., C. D. Allis, and E. Bernstein (2007). "Epigenetics: A landscape takes shape." *Cell* 128(4): 635–8.

Goldhaber, D. E. (1986). *Life-Span Human Development.* New York: Harcourt Brace Jovanovich.

Goldhaber, D. E. (2000). *Theories of Human Development: Integrative Perspectives.* New York: McGraw-Hill.

Goldsmith, H. H. (1993). "Nature-nurture issues in behavioral genetic context: Overcoming barriers to communication." In *Nature, Nurture & Psychology,* edited by R. Plomin and G. E. McClearn. Washington, DC: American Psychological Association, pp.325–340.

Goldsmith, H. H. (1994). "The behavior genetic approach to development and experience: Contexts and constraints." *SRCD Newsletter* 1(6): 10–11.

Gottesman, I. (1963). "Genetic aspects of intelligent behavior." In *Handbook of Mental Deficiency: Psychological Theory and Research,* edited by N. R. Ellis. New York: McGraw-Hill, pp. 253–96.

Gottesman, I. and D. R. Hanson (2005). "Human development: Biological and genetic processes." *Annual Review of Psychology* 56: 263–286.

Gottlieb, G. (1983). "The psychobiological approach to developmental issues." In *Handbook of Child Psychology,* edited by P. H. Mussen. New York: Wiley, Vol. 2, pp. 1–26.

Gottlieb, G. (1991). "Experiential canalization of behavior development: Theory." *Developmental Psychology* 27(1): 4–13.

Gottlieb, G. (1992). *Individual Development and Evolution.* New York: Oxford University Press.

Gottlieb, G. (1995). "Some conceptual deficiencies in 'developmental' behavior genetics." *Human Development* 38: 131–41.

Gottlieb, G. (1997). *Synthesizing Nature-Nurture.* Mahwah, NJ: Lawrence Erlbaum.

Gottlieb, G. (2002). "Developmental-behavioral initiation of evolutionary change." *Psychological Review* 109: 211–18.

Gottlieb, G. (2003). "On making behavioral genetics truly developmental." *Human Development* 46(6): 337–355.

Gottlieb, G. (2007). "Probabilistic epigenesis." *Developmental Science* 10(1): 1–11.

Gould, S. (1991). "Exaptation: A crucial tool for an evolutionary psychology." *Journal of Social Issues* 47(3): 43–65.

Gould, S. and R. Lewontin (1979). "The spandrels of San Marco and the Panglossian paradigm: A critique of the adaptationist programme." *Proceedings of the Royal Society of London. Series B* 205(1161): 581–98.

Gould, S. J. (1981). *The Mismeasure of Man.* New York: Norton.

Gould, S. J. (1996). *The Mismeasure of Man* (rev. ed.). New York: Norton.

Greenberg, G., T. Partridge, V. Mosack, and C. Lambdin (2006). "Psychology is a developmental science." *International Journal of Comparative Psychology* 19: 185–205.

Greenspan, R. J. (2001). "The flexible genome." *Nature Reviews Genetics* 2(5): 383–7.

Griffiths, P. E. and R. D. Knight (1998). "What is the developmentalist challenge?" *Philosophy of Science* 65(2): 253–8.

Griffiths, P. E. and J. Tabery (2008). "Behavioral genetics and development: Historical and conceptual causes of controversy." *New Ideas in Psychology* 26(3): 332–52.
Grotuss, J., D. Bjorklund, and A. Csinady (2007). "Evolutionary developmental psychology: Developing human nature." *Acta Psychologica Sinica* 39(3): 439–53.
Guo, S. (2000). "Gene-environment interactions and the mapping of complex traits: Some statistical models and their implications." *Human Heredity* 50: 286–303.
Hall, B. (2003). "Evo-Devo: Evolutionary developmental mechanisms." *International Journal of Developmental Biology* 47(7/8): 491–6.
Harris, J. (1999). *The Nurture Assumption: Why Children Turn out the Way They Do.* New York: Simon & Schuster.
Herrnstein, R. (1973). *IQ in the Meritocracy.* Boston, MA: Atlantic Monthly Press.
Herrnstein, R. J. (1994). *The Bell Curve : Intelligence and Class Structure in American Life.* New York: Free Press.
Horowitz, F. D. (1992). "John B. Watson's legacy: Learning and environment." *Developmental Psychology* 28(3): 360–7.
Horowitz, F. D. (1993). "The need for a comprehensive new environmentalism." In *Nature, Nurture & Psychology*, edited by R. Plomin and G. E. McClearn. Washington, DC: American Psychological Association, pp. 341–54.
Hull, C. L. (1943). *Principles of Behavior.* New York: Appleton-Century-Crofts.
Hunt, E. (1997). "Nature vs nurture: The feeling of *vuja de*." In *Intelligence, Heredity, and Environment*, edited by R. J. Sternberg and E. Grigorenko. Cambridge: Cambridge University Press, pp. 531–51.
Hunt, J. McV. (1961). *Intelligence and Experience.* New York: Ronald Press.
Hunt, J. McV. (1969). "Has compensatory education failed? Has it been attempted?" *Harvard Educational Review* 39(2): 278–300.
Jablonka, E. (2007). "The developmental construction of heredity." *Developmental Psychobiology* 49: 808–17.
Jablonka, E. and M. Lamb (2002). "The changing concept of epigenetics." *Annals of the New York Academy of Sciences* 981: 82–96.
Jacobson, K. (2009). "Considering interactions between genes, environments, biology, and social context." *Psychological Science Agenda* 23(4). Available at (http://www.apa.org/science/about/psa/2009/04/sci-brief.aspx).
Jacquard, A. (1983). "Heritability: One word, three concepts." *Biometrics* 39: 465–77.
Jensen, A. R. (1969). "How much can we boost IQ and scholastic achievement?" *Harvard Educational Review* 39: 1–123.
Jensen, A. R. (1997). "The puzzle of non-shared variance." In *Intelligence, Heredity, and Environment*, edited by R. J. Sternberg and E. Grigorenko. Cambridge: Cambridge University Press, pp. 42–89.
Johnston, T. and L. Edwards (2002). "Genes, interactions, and the development of behavior." *Psychological Review* 109(1): 26–34.
Johnston, T. D. (1987). "The persistence of dichotomies in the study of behavioral development." *Developmental Review* 7(2): 149–82.
Kagan, J. (1969). "Inadequate evidence and illogical conclusions." *Harvard Educational Review* 39(2): 274–7.
Kagan, J. (2001). "Biological constraint, cultural variety, and psychological structures." *Annals of the New York Academy of Sciences* 935(1): 177–90.
Kagan, J. (2006). *An Argument for Mind.* New Haven, CT: Yale University Press.

Kamin, L. J. (1974). *The Science and Politics of I.Q.* Potomac, MD: Lawrence Erlbaum.
Keller, E. F. (2010). *The Mirage of a Space Between Nature and Nurture.* Durham, NC: Duke University Press.
Kingsbury, K. (2009). "Why parents (still) don't matter." *Time,* Feb. 4. Available at (http://www.time.com/time/health/article/0,8599,1881384,00.html).
Kohn, M. (2006). *Change and Stability: A Cross-National Analysis of Social Structure and Personality.* Boulder, CO: Paradigm Press.
Kohn, M., A. Naoi, C. Schoenbach, C. Schooler, and K. M. Slomczynski (1990). "Position in the class structure and psychological functioning in the United States, Japan, and Poland." *American Journal of Sociology* 95(4): 964–1008.
Kovas, Y., C. Haworth, P. S. Dale, and R. Plomin (2007). "The genetic and environmental origins of learning abilities and disabilities in the early school years." *Monographs of the Society for Research in Child Development* 72(3): 1–144.
Krebs, D. (2003). "Fictions and facts about evolutionary approaches to human behavior: Comment on Lickliter and Honeycutt (2003)." *Psychological Bulletin* 129: 842–7.
Kuo, Z. Y. (1924). "A psychology without heredity." *Psychological Review* 31(6): 427–48.
Kuo, Z. Y. (1929). "The net result of the anti-heredity movement in psychology." *Psychological Review* 36(3): 181–99.
Kuo, Z. Y. (1976). *The Dynamics of Behavioral Development: An Epigenetic View.* New York: Plenum.
Lamm, E. and E. Jablonka (2008). "The nurture of nature: Hereditary plasticity in evolution." *Philosophical Psychology* 21(3): 305–19.
Leahy, A. M. (1935). "Nature-nurture and intelligence." *Genetic Psychology Monograph* 17(4): 236–308.
Lee, R. and R. Daly (1999). *The Cambridge Encyclopedia of Hunters and Gatherers.* Cambridge: Cambridge University Press.
Lehrer, J. (2009) "Do Parents Matter?" *Scientific American Online,* April 9. Available at (http://www.scientificamerican.com/article.cfm?id=parents-peers-children).
Lerner, R. (2006). "Resilience as an attribute of the developmental system." *Annals of the New York Academy of Sciences* 1094(1): 40–51.
Lewis, M. D. (2000). "The promise of dynamic systems approaches for an integrated account of human development." *Child Development* 71: 36–43.
Lewkowicz, D. J. (2011). "The biological implausibility of the nature-nurture dichotomy and what it means for the study of infancy." *Infancy* 16(4): 331–67.
Lewontin, R. C. (2006). "The analysis of variance and the analysis of causes." *International Journal of Epidemiology* 35: 52–55.
Lickliter, R. (2008). "The growth of developmental thought: Implications for a new evolutionary psychology." *New Ideas in Psychology* 26(3): 353–69.
Lickliter, R. (2009). "The fallacy of partitioning: Epigenetics' validation of the organism-environment system." *Ecological Psychology* 21(2): 138–46.
Lickliter, R. and H. Honeycutt (2003). "Developmental dynamics: Toward a biologically plausible evolutionary psychology." *Psychological Bulletin* 129(6): 819–35.
Lickliter, R. and S. Schneider (2006). "Role of development in evolutionary change: A view from comparative psychology." *International Journal of Comparative Psychology* 19(2): 151–69.

Logan, C. A. and T. D. Johnston (2007). "Synthesis and separation in the history of 'nature' and 'nurture.' " *Developmental Psychobiology.* (Special Issue: *Gilbert Gottlieb's legacy: Probabilistic epigenesis and the development of individual and species)* 49(8): 758–69.

Lombardo, P. (1996). "Medicine, eugenics, and the Supreme Court: From coercive sterilization to reproductive freedom." *Journal of Contemporary Health Law and Policy* 13: 1–13.

Lorenz, K. (1965). *Evolution and Modification of Behavior.* Chicago, IL: University of Chicago Press.

Lupien, S., S. King, M. J. Meaney, and B. S. McEwen (2000). "Child's stress hormone levels correlate with mother's socioeconomic status and depressive state." *Biological Psychiatry* 48(10): 976–980.

Magnusson, D. and R. B. Cairns (1996). "Developmental science: Toward a unified framework." In *Developmental Science*, edited by R. B. Cairns, G. H. J. Elder, Jr., and E. J. Costello. Cambridge: Cambridge University Press, pp. 7–29.

Marlowe, F. (2005). "Hunter gatherers and human evolution." *Evolutionary Anthropology: Issues, News, and Reviews* 14(2): 54–67.

Mather, K. (1964). "R.A. Fisher's work in genetics." *Biometrics* 20(2): 330–42.

McCall, R. B. (1981). "Nature-nurture and the two realms of development: A proposed integration with respect to mental development." *Child Development* 52(1): 1–12.

McCandless, B. R. and C. C. Spiker (1956). "Experimental research in child psychology." *Child Development* 27(1): 75–80.

McCartney, K. and D. Berry (2009). "Whether the environment matters more for children in poverty." In *Experience and Development*, edited by K. McCartney and R. A. Weinberg. New York: Psychology Press, pp. 99–124.

McClelland, G. H. and C. M. Judd (1993). "Statistical difficulties of detecting interactions and moderator effects." *Psychological Bulletin* 114(2): 376–90.

McHughen, S. A., P. F. Rodriguez, J. A. Kleim, E. D. Kleim, L. M. Crespo, V. Procaccio, and S. C. Cramer (2010). "BDNF Val^{66}Met polymorphism influences motor system function in the human brain." *Cerebral Cortex* 20(5): 1254–1262.

Meaney, M. (2004). "The nature of nurture: Maternal effects and chromatin remodeling." In *Essays in Social Neuroscience*, edited by J. T. Cacioppo and G. G. Berntson. Cambridge, MA: MIT Press, pp. 1–14.

Meaney, M. J. (2001). "Nature, nurture, and the disunity of knowledge." *Annals of the New York Academy of Sciences* 935(1): 50–61.

Meaney, M. J. (2010). "Epigenetics and the biological definition of gene × environment interactions." *Child Development* 81(1): 41–79.

Meaney, M. and M. Szyf (2005). "Maternal care as a model for experience-dependent chromatin plasticity?" *Trendss in Neurosciences* 28(9): 456–63.

Moffitt, T., A. Caspi, H. Harrington, and B. J. Milne (2002). "Males on the life-course-persistent and adolescence-limited antisocial pathways: Follow-up at age 26 years." *Development and Psychopathology* 14(1): 179–207.

Molenaar, P. (2007). "On the implications of the classical ergodic theorems: Analysis of developmental processes has to focus on intra individual variation." *Developmental Psychobiology* 50(1): 60–9.

Murray, C. (2008a). Down with the four-year college degree. *Cato unbound.* Available at (http://www.cato-unbound.org/2008/10/06/charles-murray/down-with-the-four-year-college-degree/). (Accessed Oct. 6.)

Murray, C. (2008b). *Real Education: Four Simple Truths for Bringing America's Schools Back to Reality.* New York: Three Rivers Press.

Murray, C. (2010). "The tea party warns of a new elite. They're right." *Washington Post,* Oct. 25. Available at (http://live.washingtonpost.com/outlook:-the-tea-party-warns-of-a-new-elite-they're-right-.html).

Neff, W. (1938). "Socioeconomic status and intelligence: A critical survey." *Psychological Bulletin* 35(10): 727–757.

Newson, L., P. J. Richerson, and R. Boyd (2007). "Cultural evolution and the shaping of cultural diversity." In *Handbook of Cultural Psychology,* edited by S. Kitayama and D. Cohen. New York: Guilford Press, pp. 454–76.

Oliver, B. and R. Plomin (2007). "Twins' Early Development Study (TEDS): A multivariate, longitudinal genetic investigation of language, cognition and behavior problems from childhood through adolescence." *Twin Research and Human Genetics* 10(1): 96–105.

Overton, W. (2006). Developmental psychology: Philosophy, concepts, methodology. In *Handbook of Child Psychology,* edited by W. Damon and R. Lerner. New York: Wiley, Vol. 1, pp. 18–88.

Overton, W. (2010). "Letter to the editor: To have a revolution you need a revolutionary metatheory." *Human Development* 53(3): 105–107.

Overton, W. F. (1984). "World views and their influence on psychological theory and research: Kuhn-Lakatos-Laudan." In *Advances in Child Development and Behavior,* edited by H. W. Reese. San Diego, CA: Academic, Vol. 18, pp. 194–226.

Overton, W. F. (2003). "Metatheoretical features of behavior genetics and development." *Human Development* 46: 356–361.

Oyama, S. (1985). *The Ontogeny of Information.* Cambridge: Cambridge University Press.

Partridge, T. and G. Greenberg (2010). "Contemporary ideas in physics and biology in Gottlieb's psychology." In *Handbook of Developmental Science, Behavior, and Genetics,* edited by K. E. Hood, C. T. Halpern, G. Greenberg, and R. Lerner. Available from Wiley Online Library, pp. 166–202.

Pastore, N. (1949). *The Nature-Nurture Controversy.* New York: Garland.

Paton, G. (2007). "Children learn most from peers not parents." *The Telegraph.* Available at (http://www.telegraph.co.uk/news/uknews/1549711/Children-learn-most-from-peers-not-parents.html).

Pepper, S. C. (1961). *World Hypotheses: A Study in Evidence.* Los Angeles, CA: University of California Press.

Pertea, M. and S. L. Salzberg (2010). "Between a chicken and a grape: Estimating the number of human genes." *Genome Biology* 11: 206.

Peters, J., T. Shackelford, and D. M. Buss (2002). "Understanding domestic violence against women: Using evolutionary psychology to extend the feminist functional analysis." *Violence and Victims* 17(2): 255–64.

Pinker, S. (2002). *The Blank Slate: The Modern Denial of Human Nature.* London: Allen Lane.

Pinker, S. (2004). "Why nature & nurture won't go away." *Daedalus* 133(4): 5–17.

Pinker, S. (2005). "So how does the mind work?" *Mind & Language* 20(1): 1–24.

Pinker, S. (2006). *The Evolutionary Psychology of Religion.* Madison, WI: Freedom from Religion Foundation, pp. 1–11.

Pinker, S. (2011). *The Better Angels of Our Nature: Why Violence Has Declined.* New York: Viking.

Pinker, S. and E. S. Spelke (2005). "The science of gender and science: Pinker vs. Spelke A Debate." Available at *Edge: The Third Culture* website: http://edge.org/3rd_culture/debate05/debate05_index.html.

Plomin, R. (1990a). *Nature and Nurture: An Introduction to Human Behavioral Genetics.* Pacific Grove, CA: Brooks/Cole.

Plomin, R. (1990b). "Trying to shoot the messenger for his message." *Behavioral and Brain Sciences* 13: 144.

Plomin, R. (1994). *Genetics and Experience: The Interplay Between Nature and Nurture.* Thousand Oaks, CA: Sage.

Plomin, R. (2007). "Genetics and developmental psychology." *Merrill-Palmer Quarterly* 50(3): 341–52.

Plomin, R. (2009). "The nature of nurture." In *Experience and Development*, edited by K. McCartney and R. A. Weinberg. New York: Psychology Press, pp. 61–80.

Plomin, R. and K. Asbury (2005). "Nature and nurture: genetic and environmental influences on behavior." *The Annals of the American Academy of Political and Social Science* 600: 86–98.

Plomin, R. and D. Daniels (1987). "Why are children in the same family so different from one another?" *Behavioral and Brain Sciences* 10: 1–59.

Plomin, R. and J. C. DeFries (1983). "The Colorado Adoption Study." *Child Development* 54(2): 276–290.

Plomin, R. et al. (1993). "Genetic change and continuity from fourteen to twenty months: The MacArthur Longitudinal Twin Study." *Child Development* 64(5): 1354–1376.

Plomin, R., Y. Kovas, and C. M. Haworth (2007). "Generalist genes: Genetic links between brain, mind, and education." *Mind, Brain, and Education* 1(1): 11–19.

Plomin, R., M. J. Owen, and P. McGuffin (1994a). "The genetic basis of complex human behaviors." *Science* 264: 1733–1739.

Plomin, R., D. Reiss, E. Hetherington, and G. W. Howe (1994b). "Nature and nurture: Genetic contributions to measures of the family environment." *Developmental Psychology* 30(1): 32–43.

Postman, N. (1982). *The Disappearance of Childhood.* New York: Vintage Books.

Pray, L. (2004). "Epigenetics: Genome, meet your environment." *The Scientist* 18(13): 14–20.

Rao, C. R. (1992). "R.A. Fisher: The founder of modern statistics." *Statistical Science* 7(1): 34–48.

Reese, H. W. and W. F. Overton (1970). "Models of development and theories of development." In *Life-Span Developmental Psychology*, edited by L. R. Goulet and P. B. Baltes. New York: Academic, pp. 116–45.

Reiss, D. (1993). "Genes and the environment: Siblings and synthesis." In *Nature, Nurture & Psychology*, edited by R. Plomin and G. E. McClearn. Washington, DC: American Psychological Association, pp. 417–32.

Richardson, K. and S. Norgate (2005). "The equal environments assumption of classical twin studies may not hold." *British Journal of Educational Psychology* 75: 339–50.
Richardson, K. and S. H. Norgate (2006). "A critical analysis of IQ studies of adopted children." *Human Development* 49(6): 319–35.
Rose, H. and S. Rose, Eds. (2000). *Alas, Poor Darwin*, London: Jonathan Cape.
Rowe, D. C. (2005). "Under the skin: On the impartial treatment of genetic and racial differences." *American Psychologist* 60(1): 60–70.
Rowe, D. C. and J. L. Rodgers (1997). "Poverty and behavior: Are environmental measures nature and nurture?" *Developmental Review* 17(3): 358–75.
Rutter, M. (2002). "Nature, nurture, and development: From evangelism through science toward policy and practice." *Child Development* 73(1): 1–21.
Rutter, M. (2007). "Gene-environment interdependence." *Developmental Science* 10(1): 12–18.
Rutter, M., A. Pickles, R. Murray, and L. Eaves (2001). "Testing hypotheses on specific environmental causal effects on behavior." *Psychological Bulletin* 127(3): 291–324.
Ryan, C. and C. Jetha (2010). *Sex at Dawn: The Prehistoric Origins of Modern Sexuality.* New York: Harper.
Sarason, S. (1973). "Jewishness, blackishness, and the nature-nurture controversy." *American Psychologist* 28: 962–971.
Scarr, S. (1991). "Theoretical issues in investigating intellectual plasticity." In *Plasticity of Development*, edited by S. E. Brauth, W. S. Hall, and R. J. Dooling. Cambridge, MA: MIT Press, pp. 57–72.
Scarr, S. (1992). "Developmental theories for the 1990s: Developmental and individual differences." *Child Development* 63(1): 1–19.
Scarr, S. (1993). "Biological and cultural diversity: The legacy of Darwin for development." *Child Development* 64(5): 1333–53.
Scarr, S. (1996). "How people make their own environments: Implications for parents and policy makers." *Psychology, Public Policy, and Law* 2(2): 204–28.
Scarr, S. (1997). "Behavior-genetic and socialization theories of intelligence: Truce and reconciliation." In *Intelligence*, edited by R. J. Sternberg and E. Grigorenko. Cambridge: Cambridge University Press, pp. 3–42.
Scarr, S. (2009). "Epilogue." In *Experience and Development*, edited by K. McCartney and R. A. Weinberg. New York: Psychology Press, pp. 253–68.
Scarr, S. and K. McCartney (1983). "How people make their own environments: A theory of genotype-environment effects." *Child Development* 54(2): 424–35.
Scarr, S. and A. Ricciuti (1991). "What effects do parents have on their children." In *Directors of Development: Influences on the Development of Children's Thinking*, edited by L. Okagaki and R. J. Sternberg. Hillsdale, NJ: Lawrence Erlbaum, pp. 3–23.
Scarr, S. and R. A. Weinberg (1983). "The Minnesota Adoption Studies: Genetic differences and malleability." *Child Development* 54(2): 260–8.
Scarr, S. and R. A. Weinberg (1994). "Educational and occupational achievements of brothers and sisters in adoptive and biologically related families." *Behavior Genetics* 24(4): 301–25.
Schmitt, D. P. (2008). "Evolutionary psychology research methods." In *Foundations of Evolutionary Psychology*, edited by C. Crawford and D. Krebs. New York: Lawrence Erlbaum, pp. 215–27.

Schweinhart, L., J. Montie, Z. Xiang, W. S. Barnett, C. R. Belfield, and M. Nores (2005). *The HighScope Perry Preschool Study Through Age 40*. Ypsilanti, MI: HighScope Press.

Segerstrale, U. (2000). *Defenders of the Truth: The Battle for Science in the Sociobiology Debate and Beyond*. New York: Oxford University Press.

Shonkoff, J. P. (2010). "Building a new biodevelopmental framework to guide the future of early childhood policy." *Child Development* 81(1): 357–67.

Simpson, T., P. Carruthers, S. Laurence, and S. Stich (2005). "Introduction: Nativism past and present." In *The Innate Mind: Structure and Contents*, edited by P. Carruthers, S. Laurence, and S. Stich. Oxford, UK: Oxford University Press, pp. 3–19.

Skinner, B. F. (1938). *The Behavior of Organisms: An Experimental Analysis*. New York: Appleton-Century-Crofts.

Skinner, B. F. (1953). *Science and Human Behavior*. New York: Free Press.

Spelke, E. S. (1998). "Nativism, empiricism, and the origins of knowledge." *Infant Behavior and Development* 21(2): 181–200.

Spelke, E. S. and K. D. Kinzler (2009). "Innateness, learning, and rationality." *Child Development Perspectives* 3(2): 96–8.

Spence, K. W. (1956). *Behavior Theory and Conditioning*. New Haven, CT: Yale University Press.

Spiker, C. C. (1989). "Cognitive development: Mentalistic or behavioristic?" In *Advances in Child Development and Behavior*, edited by H. W. Reese. New York: Academic, Vol. 21, pp. 73–90.

Sternberg, R. J., E. L. Grigorenko, and K. K. Kidd (2005). "Intelligence, race, and genetics." *American Psychologist* 60(1): 46–59.

Stevenson, H. W. (1983). "How children learn – The quest for a theory." In *Handbook of Child Psychology*, edited by P. H. Mussen. New York: Wiley, Vol. 1, pp. 213–36.

Stotz, K. (2008). "The ingredients for a postgenomic synthesis of nature and nurture." *Philosophical Psychology* 21(3): 359–81.

Szyf, M., P. McGowan, and M. J. Meaney (2008). "The social environment and the epigenome." *Environmental and Molecular Mutagenesis* 49(1): 46–60.

Thelen, E. (1995). "Motor development: A new synthesis." *American Psychologist* 50(2): 79–95.

Thornhill, R. and C. T. Palmer (2004). "Evolutionary life history perspective on rape." In *Evolutionary Psychology, Public Policy and Personal Decisions*, edited by C. Crawford and C. Salmon. Mahwah, NJ: Lawrence Erlbaum, pp. 249–74.

Tooby, J. and L. Cosmides (1997). Letter to the Editor. *The New York Times Review of Books*, July 7.

Tooby, J. and L. Cosmides (2005). "Conceptual foundations of evolutionary psychology." In *Handbook of Evolutionary Psychology*, edited by D. M. Buss. Hoboken, NJ: Wiley, pp. 5–67.

Tooby, J., L. Cosmides, and H. C. Barrett (2003). "The second law of thermodynamics is the first law of psychology: Evolutionary developmental psychology and the theory of tandem, coordinated inheritances: Comment on Lickliter and Honeycutt (2003)." *Psychological Bulletin* 129: 858–65.

Tooby, J., L. Cosmides, and H.C. Barrett (2005). "Resolving the debate on innate ideas." In *The Innate Mind*, edited by P. Carruthers, S. Laurence, and S. Stich. Oxford, UK: Oxford University Press, pp. 305–37.

Trivers, R. (1972). "Parental investment and sexual selection." In *Sexual Selection and the Descent of Man*, edited by B. Campbell. New York: Aldine de Gruyter, pp. 136–79.

Tudge, J., I. Mokrova, B. E. Hatfield, and R. B. Karnik (2009). "Uses and isuses of Bronfenbrenner's Bioecological Theory of Human Development." *Journal of Family Theory & Review* 1(4): 198–210.

Tudge, J., S. Putnam, and J. Valsiner (1996). "Culture and cognition in developmental perspective." In *Developmental Science*, edited by R. B. Cairns, G. H. J. Elder Jr., and E. J. Costello. Cambridge: Cambridge University Press, pp. 190–222.

Turkheimer, E. (2000). "Three laws of behavior genetics and what they mean." *Current Directions in Psychological Science* 9(5): 160–4.

Turkheimer, E. (2004). "Spinach and ice cream: Why social science is so difficult." In *Behavior Genetics Principles: Perspectives in Development, Personality, and Psychopathology*, edited by L. F. DiLalla. Washington, DC: American Psychological Association, pp. 161–89.

Turkheimer, E. and M. Waldron (2000). "Nonshared environment: A theoretical, methodological, and quantitative review." *Psychological Bulletin* 126(1): 78–108.

Turkheimer, E., K. P. Harden, B. D'Onofrio, and I. I. Gottesman (2009). "The Scarr–Rowe interaction between measured socioeconomic status and the heritability of cognitive ability." In *Experience and Development*, edited by K. McCartney and R. A. Weinberg. New York: Psychology Press, pp. 81–98.

van der Maas, H. L. J. and B. Hopkins (1998). "Developmental transitions: So what's new?" *British Journal of Developmental Psychology* 16: 1–13.

van der Maas, H. L. J. and P. C. M. Molenaar (1992). "Stagewise cognitive development: An application of catastrophe theory." *Psychological Review* 99(3): 395–417.

van Geert, P. (1998). "We almost had a great future behind us: The contribution of non linear dynamics to developmental science in the making." *Developmental Science* 1(1): 143–59.

Vreeke, G. J. (2000). "Nature, nurture and the future of the analysis of variance." *Human Development* 43: 32–45.

Wachs, T. D. (1993). "The nature-nurture gap: What we have here is a failure to communicate." In *Nature, Nurture & Psychology*, edited by R. Plomin and G. H. McClelland. Washington, DC: American Psychological Association, pp. 375–90.

Wade, N. (2009). The evolution of the God gene. *The New York Times*, Nov. 14. Available at (http://www.nytimes.com/2009/11/15/weekinreview/12wade.html).

Wahlsten, D. (1990). "Insensitivity of the analysis of variance to heredity-environment interaction." *Behavioral and Brain Sciences* 13: 109–61.

Wahlsten, D. (2000). "Analysis of variance in the service of interactionism." *Human Development* 43: 46–50.

Watson, J. B. (1930). *Behaviorism*. Chicago, IL: University of Chicago Press.

Watters, E. (2006). "DNA is not destiny." *Discover Magazine*, Nov. 22. Available at (http://discovermagazine.com/2006/nov/cover/article_view?b_start:int=2&-C=).

Weikart, D. P., D. Deloria, S. Lawser, and R. Wiegarink (1970). "Longitudinal results of the Ypsilanti Perry Preschool Project." Ypsilanti, MI: HighScope Educational Research Foundation, Monograph Series.

Weizmann, F. (1971). "Correlational statistics and the nature-nurture problem." *Science* 171(3971): 589.

West, M. and A. King (1987). "Settling nature and nurture into an ontogenetic niche." *Developmental Psychobiology* 20(5): 549–62.

West-Eberhard, M. (1998). "Evolution in the light of developmental and cell biology, and vice versa." *Proceedings of the National Academy of Sciences of the United States of America* 95(15): 8417–19.

West-Eberhard, M. (2005). "Developmental plasticity and the origin of species differences." *Proceedings of the National Academy of Sciences* 102(Suppl 1): 6543–49.

Whitfield, K. E. and G. McClearn (2005). "Genes, environment, and race." *American Psychologist* 60(1): 104–14.

Wiebe, S., K. Espy, C. Stopp, J. Respass, P. Stewart, T. R. Jameson, D. G. Gilbert, and J. I Huggenvik (2009). "Gene-environment interactions across development: Exploring DRD2 genotype and prenatal smoking effects on self-regulation." *Developmental Psychology* 45(1): 31–44.

Wohlwill, J. F. (1973). *The Study of Behavioral Development*. New York: Academic.

Wong, A., I. Gottesman, et al. (2005). "Phenotypic differences in genetically identical organisms: The epigenetic perspective." *Human Molecular Genetics* 14(suppl 1): R11–R18.

Yates, F. (1964). "Sir Ronald Fisher and the design of experiments." *Biometrics* 20(2): 307–21.

Zernike, K. (2011). "Fast-tracking to kindergarten?" *New York Times*. May 15, 2011.

Index